Edge Computing and Capability-Oriented Architecture

Edge Computing and Capability-Oriented Architecture

Haishi Bai and Boris Scholl

CRC Press
Taylor & Francis Group
Boca Raton London New York

CRC Press is an imprint of the
Taylor & Francis Group, an **informa** business

First edition published 2022
by CRC Press
6000 Broken Sound Parkway NW, Suite 300, Boca Raton, FL 33487-2742
and by CRC Press

2 Park Square, Milton Park, Abingdon, Oxon, OX14 4RN

© 2022 Taylor & Francis Group, LLC

CRC Press is an imprint of Taylor & Francis Group, LLC

ISBN: 9780367549817 (hbk)
ISBN: 9781032046723 (pbk)
ISBN: 9781003091455 (ebk)

Typeset in Sabon
by KnowledgeWorks Global Ltd.

Contents

Foreword

Edge computing is a fast-evolving area, and with advances in 5G, connectivity, hardware accelerators for computing, networking and AI plus with regulatory aspects of data, its sovereignty and data privacy, it is becoming a critical part of any enterprise and business infrastructure.

No surprise that hyperscale cloud providers are fueling the rise of edge computing by bringing their services ever closer to the customers, where we have seen a rising interest in edge computing. Many consider edge computing as some form of cloud computing, but there are fundamental differences due to the heterogeneous nature of edge environments, devices and networks.

That said it is not always easy to develop the right architecture and adopt the right technology stack for the edge.

This book is a much-needed resource to address this need and to be a source of knowledge in this area. This book is special because it is written by people who are working on that very subject on a daily basis, Haishi Bai and Boris Scholl.

The first time I spoke with Haishi, it was clear that he is a brilliant software architect with a good sense of business. We discussed how edge is becoming an integral part of connected products and enterprises, and how the right architecture is essential in delivering its value.

Since then, we have had many more conversations about edge computing, how to build technology stack for it, how to scale it and specifically how to architect for it.

Of all things I have discussed with Haishi, the most thought-provoking might be around building a seamlessly scaling platform for sensorized edge AI. He showed me new angles of how we can think about scaling edge AI by tapping into install-base of existing devices and runtimes. I clearly recall that conversation and how insightful it was to me.

As a Partner Product Architect in our Cloud & AI working closely with our most important customers on strategies for bringing our cloud technologies and patterns to all form factors of edge environments, Boris's experience is the perfect addition to Haishi's experience. The two are no strangers and work closely together to come up with approaches to tackle the most difficult problems.

Haishi's ideas are practical and at the same time they provide depth combined with Boris's customer knowledge and strategic thinking in that area. I'm glad they are sharing their insights.

Moe Tanabian
Vice President, GM
Azure AI/Edge

Section I

Edge computing software
fundamentals

Chapter 1

Edge computing fundamentals

The year 2020 marked the 20th anniversary of a continuous human presence in space. At 8:01 PM Pacific Time on November 16, 2020, the SpaceX Crew Dragon capsule successfully docked at International Space Station and sent four astronauts to join three astronauts who had already been working on the space lab floating 400 kilometers above the Earth.

The docking process was totally autonomous – computers were able to perfectly align and link the two space bodies moving at 7.66 kilometers per second. Although the astronauts could step in and take over the process as needed, they sat comfortably in their seats and watched the computer do its magic, along with millions of viewers on the Internet around the world. The spacecraft guided itself to the docking port and secured itself without any human intervention – it acquired environmental information and made decisions on the spot. This kind of *in-context computation* is the focus of this book – edge computing.

For the last decade, most IT discussions were centered around cloud computing. But recently the term edge computing has gained a lot of attraction. But what exactly is edge computing, why are more and more people talking about it, how is it different than cloud computing and what are the things one needs to consider when designing and building applications to run on the edge? In this chapter, we will cover fundamentals of edge computing, driving forces for edge computing and edge computing characteristics. We intentionally skip the regular "motivational speeches" that discuss how many billions of devices will be connected in 10 years and how many dollars will be generated. As you are already reading this book, we assume you've got enough motivation to learn more about edge computing. We also want to clarify that this book focuses on software architecture and design patterns. This is not a book that teaches you how to connect a Raspberry Pi to cloud, nor does it introduce how 5G works (well, we'll talk about 5G, but just enough for our discussions). Our goal in this book is to help you navigate through the complex edge computing landscape while keeping a clear mind of how things fit and work together, and to equip you with common design patterns in your toolbox for your own edge computing projects.

Before that, we need to first define what exactly is edge computing.

1.1 FINDING THE EDGE

It's easy to identify where cloud is – cloud resides in the huge datacenters managed by the few cloud platform providers. Cloud compute resources are centrally managed, customer-neutral compute resources that can be leased out to customers or be used to host services (SaaS) that will be consumed by subscribers.

Defining edge is a bit tricky, because when examined from different perspective, the concept of edge is often implicitly associated with different scopes. Some define edge as *"edge of the cloud"*. This is a very restrictive view and limits edge computing scenarios to connected scenarios, which are just a portion of the edge computing landscape. On the other hand, some define edge as *"everything between data sources and cloud"*. This is a very loose definition that leads to many confusions, as too many things can be fit into this very generic bucket.

In this section, we will examine a few existing models of defining edge computing. Then, in the next section, we will present our definition of edge computing.

1.1.1 Edge computing models

One way to split the big bucket of edge computing into more granular segments is the **cloud-fog-dew** model. *Fog computing*, which is coined by Cisco, aims to provide compute, storage and networking capabilities on decentralized, closer to client or near-user edge devices, such as network routers and on-premises servers. *Dew computing* pushes the decentralized computing future by incorporating various application scenarios (such as AI) and devices. It makes capabilities available to consumers *in context* (on device, for example), regardless of connectivity to cloud or fog computing resources. For example, a dew computing scenario can be realized by allowing devices to share compute power and exchange functionality directly over a peer-to-peer network.

Another model is the **near-far** model. This model divides edge by proximity to the cloud datacenters. Compute resources that are closer to end users belong to *far edge* (and sometimes the terminal devices are said to belong to the *extreme edge*), compute resources that are closer to cloud datacenters belong to *near edge*. The near-far model is a fuzzy model. For example, many things can be considered near edge, including content delivery network infrastructure and telcon Central Offices (CO). And an on-premises server may be considered near edge or far edge in different contexts.

One more (informally defined) model is **device-premise-telco-cloud** model (or **edge-local-network-aggregation**, when examined from the networking perspective). In this model, *devices* refer to physical end user devices such as sensors, drones and VR/AR goggles. *Premise* refers to on-site compute resources such as local area network-based servers and field gateways.

Telcon (or carrier) refers to telcon infrastructure such as towers, base stations and central offices. And finally, *cloud* refers to compute resources in cloud datacenters connected by cloud dark fiber.

All three models try to segment edge computing by location. However, location-based modeling isn't capturing the essence of edge computing. For example, a mobile phone is generally considered an edge device. However, if we took the mobile phone and placed it in a cloud datacenter, does it become a cloud computing device? Obviously not. Hence, we believe there must be a better definition and captures what edge computing is really about.

1.1.2 Defining edge computing

We define edge computing as follows:

> *Edge computing is computing in the context of where data is generated or consumed.*

This definition captures a key characteristic of edge computing: compute in context. Edge computing performs computations in the real-world contexts, responding to physical world data. And it often aims to provide instantaneous responses. In other words, edge computing is often coupled with the physical scenario in both time and space. This is fundamentally different from cloud computing, in which compute resources are decoupled from physical contexts so that they can be efficiently shared across different scenarios. Even when a cloud compute resource handles an edge scenario, the data needs to be shipped out of the physical context into the cloud. Then, the cloud compute resource performs required calculation and sends the response back to the physical context.

Edge is where the cyber world and the physical world converge. This decides the two characteristics of edge computing: *local* and *temporal*.

- **Local.** Because of the attachments to the physical world, edge computing results are meaningful only within a physical context. Logically, it's preferable that the trigger-response loop should happen locally in context instead of going through cloud, avoiding problems such as privacy concerns and long latencies. The locality also raises some challenges to normal HA designs. For example, failing over to an out-of-context device may lose the critical contextual information.
- **Temporal.** In edge computing, data often needs to be processed not only on the spot, but also within a time window – adjusting radiator settings based on yesterday's temperature reading is less useful. Other examples are real-time fallback scenarios like cars or robots. The temporal nature often demands low latency and high availability of the required services and the supporting infrastructure.

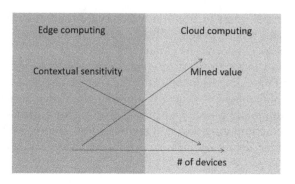

Figure 1.1 IoT scenarios.

The Internet of Things (IoT) is a special case of edge computing. A common application of IoT is to aggregate data from multiple devices and then to extract value out of the aggregated view. In such cases, compute is often conducted out-of-context (such as on cloud) over a greater time span. Figure 1.1 illustrates how IoT scenarios trade context sensitivity for potentially mined value by aggregating data from multiple devices. The diagram also shows that in the case of IoT, edge computing and cloud computing is a continuum instead of being distinct computing models. This is true for many connected edge computing scenarios. So, chances are, your edge computing projects will have an edge part and a cloud part.

This continuum brings the best of edge computing and cloud computing together. The edge part ensures relevancy to the physical context, and the cloud part generates increasing value as it aggregates more data from scattered contexts. This is a powerful combination that enables remarkably interesting scenarios. When we examine the landscape of edge computing, many scenarios fall into the IoT category, in which data is collected from devices, sent to cloud for computation and returned to device to provide feedbacks. Because of the popularity of IoT scenarios, sometimes people use IoT and edge computing as exchangeable terms. However, as we have illustrated above, IoT is a combination of edge computing and cloud computing. It cannot be used as an alternative term of edge computing.

1.2 EDGE COMPUTING CHARACTERISTICS

Although edge computing covers a great spectrum of scenarios, we can identify some unique characteristics that set edge computing apart from other computing models. This section provides a short list of such characteristics.

1.2.1 Compute in context

As we've introduced in the earlier section, a key characteristic of edge computing is computing in context. The emphasis on the linkage to the physical world is observed only in edge computing. All other computing models we are familiar with, such as client/server (C/S), browser/server (B/S), cloud computing and distributed computing, all focus on establishing compute topologies in the cyber world, without much concern with how computing is infused with the real world. Edge computing, on the other hand, is about bringing computation into real-world contexts.

Mobile phones are enormously powerful media to bring computing into real-world contexts. For example, when you play online games with your friends, on the one hand, you are using cloud computing to break the limitation of physical contexts so that you can play with your friends around the world. On the other hand, you are using edge computing to bring gaming experience into your physical contexts such as your living room and your car (hopefully not in the driving seat, of course). In this case, computation is a collaboration between cloud and edge: the local game client provides local rendering and gaming logic; while the cloud synchronizes data across different clients. When you use your mobile phone as a navigation device, the mobile phone uses satellite data to establish your physical location (through triangulation using time differences between satellite signals), and then feeds the location to navigation software that makes sure you are on the right route to your destination. Both are edge computing scenarios, as they both bring compute into physical contexts.

So far, we have mainly talked about edge models that require connections to clouds. One fallacy of distributed systems is that the network is reliable, which means that we should never assume that there is always a reliable connection to the cloud. Network failures are just one example for why one wants to have autonomous systems on the edge. Other reasons include systems that should never be exposed to a cloud and need to operate autonomously in the presence of failures. Mission critical systems such as power plants or our previous example of cars require full autonomy. This effectively means that a cloud-to-edge model is only useful to a certain degree. In the case of a car that keeps uploading driver and car telemetry to the cloud, it does not matter if it is connected, as the worst thing that can happen is that some data is not being uploaded. What's more important is that a car is fully functioning in the case of a software failure and this is what autonomy refers to.

1.2.2 Capability oriented

As mentioned before in the car example, edge computing scenarios often have the requirement of service continuity with or without connectivity to cloud. Another example is a device requiring face detection capability. It may rely on a cloud service to do face detection and fall back to a local model to provide possibly degraded but continuous service when it's disconnected from the cloud or has to run on a low power device. From this device's perspective, it needs the *capability* of detecting a face. It doesn't care if the capability is delivered as a service or a local call. Informally, we call this kind of application design capability-oriented architecture (COA). COA is a fascinating topic to dig in. For example, COA requires a different set of functionalities that are provided by common orchestrators (such as Kubernetes), including capability discovery, task delegation, multi-device synchronization and distributed scheduling. We will return to COA later in this book.

The actual capability of a device to execute a given workload is also bound to the resources available on a device. In general, we can break this down into the following categories:

- Cloud: Capabilities in the cloud are almost infinite whether it is processing, memory size of state persistence (storage). This enables *aaS platforms that in turn can run workloads.
- On-premise cloud: This is similar to cloud, but there is a boundary to which point resources can be scaled. This is determined by economics such as costs of software and hardware costs, but also space and logistics. For example, on-premises clouds can achieve resiliency and high availability within a datacenter or even two, but rarely on a global level.
- Server/VMs: In this capability category, there are no *aaS runtimes or services. The number of machines needed is usually determined by some sizing exercises and is pretty static.
- Low capability: This category is defined by low resource availability and often limited power supply. Other key differences in this category are that real time is defined in the low single-digit millisecond range as well as the use of hardware-centric programming languages.

1.2.3 Centralized management, distributed compute

A large-scale edge computing scenario often involves many devices. Hence, efficient device and workload management is a key requirement of an edge computing system. Managing many distributed, heterogeneous devices with spotty connectivity is a challenging task. A *federated control plane* that spans across multiple physical clusters is sometimes needed. An on-device *agent* is often needed to collaborate with the control plane to perform various tasks, including:

- Apply configuration updates, such as device settings, secrets and policies.
- Apply workload updates, such as acquiring and applying new workload binaries. Workload distribution can be directly from the control plane, through a separate repository (such as Docker Hub or other Docker registries) or through a manual process such as attaching a new USB stick. Workload binaries may also be distributed by a cascaded scheduling

process in which multiple levels of scheduling are performed. Workload can be distributed by an allocation process, in which the control plane decides where to place the workload. Another method of workload distribution is by an auction process, in which participating compute resources make offers to auctions conducted by the control plane.

- Establish connectivity. Although data doesn't often flow through the agent, the agent is often responsible to establish and maintain the communication channel, either through direct messaging or through a message bus.
- Report telemetry. An agent often needs to report heartbeat signals to the control plane.
- Self-registration. It's often desired that an agent can automatically bootstrap itself with the control plane to provide a *zero-touch provisioning* experience.

The control plane provides a single-pane-of-glass view of the entire system. It's responsible to drive the system toward the *desired state*. For example, a centralized control plane can allow editing the desired topology of the entire system, and make sure it's properly materialized on actual devices through mechanisms such as bulk provisioning. Each device can have a projection (often called a *device twin* or *device shadow*) that acts as an abstraction of the physical device. Users can use management user interface to update the desired states of the projection, and the control plan pushes the state to the physical device when possible.

1.2.4 Secured

Many broadband routers in your houses come with default administrator accounts and passwords, which you can easily look up on the Internet. Having a default password is convenient for maintenance – forgetting passwords is one of the top tech support requests IT companies face. Furthermore, when you have connection problems and every diagnose attempts have failed, you can always reset your router, log in using the default credentials and configure everything from scratch. However, what's convenient for you is also convenient to hackers. For example, a hacker that gets hold of your router can update the domain name system setting to direct your traffic to phishing sites. Your machines may also be injected with malware and be leveraged as compute notes to attack high-value targets.

When compute resources are centrally managed, you can use a firewall to build up a security barrier around your compute assets. Things are more complicated in an edge environment, as devices are often scattered and managed with different levels of security expertise. When trying to compromise a system, hackers always attack the weakest link in a system. And edge computing has a massive attack surface. So, any serious edge computing solutions should take security very seriously. A good practice is to assume that everything can and will be compromised. The distributed nature of edge computing makes security a challenging topic that encompasses many aspects, including (but not limited to):

- Authentication and authorization
- Device identity and secret management
- Secured connection and secured messaging
- Fraud and abnormality detection
- Tenant isolation
- Secure boot, confidential computing and secure multi-party computing
- Data protection at rest, during transition and in use
- Hardware and software integrity
- Firmware, OS and software update and patching

We'll discuss edge computing security in more detail in Chapter 2.

1.2.5 Heterogeneity

Edge computing is very segmented, with many different players at different levels, including cloud platforms, telco companies, hardware manufactures, independent software vendors, standard bodies, content providers (such as over-the-top video service providers and gaming companies), enterprises and consumers.

Various software and hardware products, standards, protocols and operators coexist in the massive edge computing ecosystem. Over the years, there have been numeric attempts to establish unified standards and protocols. However, most of them ended with adding more standards and protocols to the mix. We've also observed an *agent proliferation* on edge devices – as more systems try to plug their agents onto edge devices, there are an increasing number of device agents.

Sometimes, segmentations are intentional: a device maker may want to create an ecosystem of its own devices by ensuring interconnectivity of these devices. However, at the same time, the device maker may set up barriers for other devices to join the ecosystem to steal market shares.

Creating some sort of uniformity out of this chaos is a challenging however increasingly necessary task. As edge computing scenarios become more sophisticated and devices become more specialized, you often need to orchestrate multiple devices from different vendors to compose a complete edge computing solution. However, it's unrealistic to expect all the management systems from different vendors are compatible. And it even less realistic to expect all of the devices to adapt the same software stack (just think of tiny devices running real-time operating system and edge gateways running full Linux distributions) or communication protocol.

Is the situation really that despairing? We think there are two ways to create consistency: focusing on workflow and allowing multiple perspectives of the same logical stack.

Imagine you are doing a machine learning project. You may choose to use different neural network architectures such as ResNet and LSTM, different frameworks such as TensorFlow and PyTorch, and different hardware such as graphics processing unit (GPU) and tensor processing unit (TPU). However, at a high level, the workflow is largely the same:

1. Create a data pipeline to collect data. This may be collected sensor data, batch imported data or synthetic data.
2. Clean up data. Label data if necessary (for supervised learning).
3. Select features to be considered in the model. This is also called feature engineering.
4. Select a model architecture.
5. Train the model with data. This includes iterative process of tuning model parameters and hyper parameters and retraining.
6. Publish the model as a consumable format, such as an inference service or an inference application (that is often containerized).
7. Run inferences and monitor model performance.
8. Collect retraining data and continue with step 5.

When we described the above workflow, we didn't refer to any specific model architectures, frameworks or hardware – this is a platform-agnostic workflow. And this is our opportunity to create uniformity across different platforms. When we uplift the conversation to workflow level, we can create a consistent user experience across different platforms. Essentially, all platforms become implementation details of workflow steps. To realize this, however, we need a language that abstracts all technical details while allowing semantic meetings of workflow steps be explicitly described. When we discuss COA in Chapters 7 and 8, we'll examine how a common lexicon makes this possible.

When a device goes through its life span, it is configured and managed by multiple systems that come from different perspectives, including:

- Device provisioning
- Device management
- Workload management
- Data pipeline and monitoring
- Data analysis and reporting

The same physical device may have different presences in different systems. Without a single source of truth, it's hard to track a device going through these systems. One possible approach is to create a brand-new state store that is shared by all systems. And you can imagine different projectors being created to convert data schemas. As long as the new data store is the only store that takes writes, all other data stores can be used as read-only projections of the unified data store and become eventually consistent.

If that's impossible, you may consider picking one of the systems (such as the system that's mostly used) as the primary system and use data projections to project read-only copies of data to other participating systems. If neither is possible. You may have to figure out a complex data replication and recon- ciliation scheme that synchronizes data in an eventually consistent manor. We'll further develop this multi-perspective idea in Chapters 6–8.

Once we understand the key characteristics of edge computing, we can examine typical edge computing scenarios. Then, in Chapters 3 and 4, we'll introduce different edge computing design patterns and architectures.

1.3 EDGE COMPUTING SCENARIOS

Because of the broad range of edge computing, many scenarios can be con- sidered as edge computing scenarios. Hence, it's useful to put these scenar- ios into different categories. This section summarizes a couple of different methods to categorize edge computing workloads, starting with a decompo- sition of IoT scenarios.

1.3.1 IoT scenarios

As IoT evolves, the concept of IoT has been expanding to cover more spe- cific scenarios, including the Industrial IoT (IIoT), the Artificial Intelligence of Things (AIoT), the Internet of Heavier Things (IoHT) and other varia- tions. These categories are not mutually exclusive. Chances are, you can find many of the following aspects playing together in an IoT application.

- **IoT** (Internet of Things)

A system of interrelated computing devices. This is a generic term that applies to a broad spectrum of connected device scenarios.

- **IIoT** (Industrial Internet of Things)

IIoT is an evolution of a distributed control system (DCS) that enables high degree of automation. It emphasizes interconnectivity of sensors and indus- trial applications such as manufacturing. Physical machineries and com- puters are organized into a cyber-physical system (CPS) that, as a whole, has the capability of sensing physical contexts and make smart choices to influent the physical world.

- **AIoT** (Artificial Intelligence of Things)

Augment IoT devices with AI technologies. As the name suggests, AIoT focuses on fusing AI with IoT. Some believe that all IoT projects would be AI projects (just as some believed all IoT projects should be Big Data projects). Truth or not, the opinion reflects the reality of the IoT world now – people start to seek more immediate, localized value generation and direct feedbacks than value generation through a complete pipeline, which takes efforts to build and customize. As AI models become more approachable with managed AI services and open-sourced AI frameworks, fusing certain AI capabilities into IoT solutions becomes a common practice.

- **IoHT** (Internet of Heavier Things)

Apply IoT solutions to large and heavy equipment and facilities such as motors, generators and heavy machineries. One key difference between IoHT and other IoT scenarios is that IoHT focuses on enhancing the equipment themselves instead of collecting data from the equipment.

- **IoMT** (Internet of Medical Things)

This category focuses on applying IoT to medical and healthcare applications. Note that this category is also called Internet of Healthcare Things (IoHT). We chose IoMT here to disambiguate from IoHT. Unfortunately, IoMT may also refer to Internet of Military Things, which we won't discuss further in this book. One interesting trend in IoT is to convert devices to subscribed services. And this trend is often observed in IoMT to transfer expensive medical equipment such as MRI (magnetic resonance imaging) machines into a subscribed MRI service. You may wonder why this is a viable business model to offer an expensive MRI machine as a much cheaper, subscription-based service. Making MRI as a service allows hospitals that cannot afford purchasing an MRI machine to leverage a shared or leased MRI machine at a lower cost. This considerably broaden the MRI market, which wasn't penetrable because of the prohibitable cost of MRI machines.

We are sure that we've missed many other variations of IoT categories. Regardless, most IoT scenarios are essentially connected scenarios that connect physical devices to a compute mesh (such as cloud) to extract value.

1.3.2 Hardware as a service

The shared MRI machine case we mentioned above represents a new business model that treats a hardware as a service. In addition to allowing device makers to penetrate new markets, the model is also more attractive to investors because it offers a continuous, predicable cash flow after the devices have been deployed. And when the model is combined with modern device management such as predictive maintenance, it presents a more sustainable and profitable business. Although in the past few years, we've seen overhyped shared economy businesses such as shared bicycles, we believe many sustainable businesses can be created out of this model – such as Airbnb. We can call this model hardware a service (HaaS).

Three-dimensional (3D) printers used to be quite expensive. A small 3D printer could cost you thousands of dollars. Some saw the business opportunity and started to offer 3D printing as a service – instead of buying a 3D printer yourself, you submitted your print job to the service and a 3D model was printed and mailed to you. Online 3D printer farm represents another form of HaaS, which focuses on delivering the required capabilities (3D printing in this case) instead of physical hardware to consumers. In this mode, hardware is centrally managed, which allows better control and more opportunities for optimization. And because the printer farm operates as a shared "compute" plane, typical cloud platform methods such as dynamic failover (between jobs) and resource load balancing (such as balancing job queue lengths) can be applied.

Another example in the same category is mobile phone farms that are used to test mobile applications. A mobile application often needs to be tested on different platforms (such as macOS and Android) with different product lines (such as iPhone, iPad, Google Pixel and Samsung Galaxy), different OS releases and different hardware configurations. A comprehensive test harness is expensive to maintain. Online mobile phone farms solve this problem by renting out centrally managed mobile devices through remote technologies and test automation systems.

1.3.3 Hybrid scenarios

A hybrid scenario spans cloud and edge. What makes it different from typical IoT scenario is that in a hybrid scenario, the same computation task may get dynamically shifted between cloud and edge. We'll examine several typical hybrid application scenarios in this section, starting from bursting scenarios.

When a 3D rendering software need to render complex scenes with many polygons, dynamic lighting and semitransparent materials, GPU acceleration is a key to improve the performance. However, for the most challenging scenes, or long sequence of scenes (such as a featured movie), GPUs on the most powerful desktop become insufficient. In this case, the software can choose to "burst" rendering tasks to a cloud-based farm with powerful GPUs to collectively complete the rendering task. And with the elasticity of cloud, compute resources are dynamically acquired as needed and released once the task has finished. So, the rendering user pays for only the incurred compute time (with a bit of overhead, of course). This model of dynamically allocating and using cloud resources is called "bursting to cloud". The model is applicable to scenarios in which a large amount of compute resources needs to be allocated and released during a short period of time, such as performing big data analysis on increased traffic to a retail website.

Another important hybrid scenario is backup and disaster recovery (DR). In this scenario, cloud serves as a secured archive for periodical snapshots of on-premises systems. When certain disaster, such as fire and flood, happens and destroys the local infrastructure, the system can still be restored to the last snapshot, the frequency of which is driven by recovery point objective (RPO). Modern clouds also provide cold storage services that can archive data for years with a low cost. This is quite useful when enterprises need to retain a large amount of data for a long time for compliance reasons. Sometimes, cloud-based backup and recovery is also used as a migration mechanism that migrates a system from one physical location to a different physical location (such as moving to a different office in a different state or country).

A hybrid infrastructure also enables what's called "shadow IT". In this case, an IT department deliberately chooses to host service that is compatible with cloud-based services. For example, they may choose a local customer relation management (CRM) system that also has a cloud-based version. The IT department sets it up as a proxy to both service endpoints and route users that handle sensitive data to the private CRM and redirects other users to the public CRM, which can be subscribed at a lower cost than hosting the service itself. This arrangement satisfies the privacy and possible compliance requirements of the enterprise and offers a consistent user experience across the company.

The last hybrid scenario we'll cover in this section is to leverage the cloud network infrastructure to bridge two remote sites. Cloud datacenters have global presence, and they are connected by the world's largest dark fiber networks owned or leased by the cloud platform. An enterprise can use this dark fiber as an information highway between remote sites across the globe. Especially, they can use various virtual private network offerings to create an end-to-end secured private network across different office locations.

1.3.4 5G scenarios

A quick refresher of high school physics – electromagnetic waves can carry encoded information. Low frequency waves have long wavelength, which travels better in long distance; high frequency waves have shorter wavelength, which doesn't travel well especially around obstacles. High frequency waves can carry "denser" data hence provide more throughput. 5G uses higher frequency bands hence in principle can offer better throughput and lower latency. As cellular network technology progresses, more and more scenarios are enabled with increased bandwidth and reduced latency – from text message to video conference to VR streaming. ITU-R's IMT-2020 categories 5G scenarios into three broad buckets: eMBB, mMTC and URLLC.

- eMBB (Enhanced Mobile Broadband)

This group of scenarios features *bandwidth*. These scenarios require stable connections, extreme high bandwidth, moderate latency and high spectral efficiency. Typical scenarios including AR/VR scenarios, Ultra-HD video streaming, 360° video, multiplayer gaming, and video conferencing.

- **mMTC** (Massive Machine Type Communication)

This group of scenarios features *scale*. These scenarios often use IoT devices with extreme density, low power, low cost and low complexity. These devices are sporadically active and send small data payloads. Typical scenarios in this category include sensor networks, smart city, smart home, smart logistics, smart metering and many large-scale IoT applications.

- **URLLC** (Ultra-High Reliable and Low Latency)

This group of scenarios features *low latency* and *high reliability*. These scenarios require networks with low latency, high availability, strong security and high reliability. Typical scenarios include self-driving cars, industry automation, mission-critical applications such as smart grid.

There is an industry wide belief that 5G will have a tremendous impact on use cases listed below which fall into the aforementioned buckets:

- Industry 4.0: Often called the fourth industrial revolution is happing right now. Industry 4.0 is characterized by the fusion between software and physical infrastructure such as building, robots, etc. 5G will add a new dimension to machine-to-machine communication providing the ability to streamline processes, monitor the supply chain and environment, calibrate equipment and allow for zero touch operations through predictive maintenance.
- Smart Cities: 5G will enable connectivity and synchronization between all participating institutions, such as 911 services or public transport and systems such as cars or traffic systems. The improved connectivity results in smart decision ranging from traffic advisories all the way to energy savings.
- Healthcare: 5G and IoT will enable easy and fast sharing of large diagnostic images, some scanners can generate images the size of 1 GB, will allow monitoring and diagnostics through wearables and eventually even enable remote surgery.
- Retail and Consumers: Among other things 5G will enable close to real-time, secure AI-enabled personal shopping experiences for consumers and immediate inventory updates with automated backorders for retailers.

In the past decades, the computer science industry has spent a lot of energy to enable distributed computation over unreliable, low-bandwidth networks, such as various caching, encoding and compression methods, distributed algorithms that tolerate prolonged network partitioning and reliable messaging systems. As we are entering a world with pervasive, reliable high-bandwidth connections (with future 6G or 7G networks, for example), we can essentially treat distributed computational resources as collocated resources as the network failures won't be more frequent than other hardware failures such as disk failures or memory failures. Such a pervasive network paves the way for ubiquitous computing, which as we see it is the ultimate computing model that merges the cyber world and the physical world.

1.3.5 Edge computing in a pandemic

This section wasn't in our plan. However, as we write the text in 2020 during the global pandemic, we feel the urge to document and analyze the role edge computing played during this period. We especially want to analyze a few public health applications of edge computing and offer some thoughts on how we can use edge computing technologies to fight future pandemics.

As the pandemic progressed, many proposals were put forward to enhance video surveillance to monitor crowd behaviors to fight the spread of the virus. For example, some AI-based systems were proposed to monitor if people were maintaining safe distances in a public space. Keeping a safe distance was a very ineffective approach (as the virus could linger in the air for hours), and reinforcing the distance was impractical. However, some other systems could provide practical help, such as detecting people with fevers, finding people without masks and counting the number of people in a store. Such crowd surveillance provides critical intelligence for virus control; however, we need to protect privacy at the same time. For instance, systems based on depth cameras can efficiently mask the true identities of people.

To stop the virus from spreading, many countries and their local governments required people to stay at home as much as possible. And many companies, especially high-tech organizations, started to allow employees to work from home. This caused an unprecedented spike in remote office services such as shared document editing and video conferencing. Just to put things in the perspective – my monthly network usage was around 8TB of data. Yes, that was terra-bytes. As cloud vendors struggled to build up additional capacity to meet the demand, workers at home were trying to adjust to different working and living styles. Technological problems were easy to solve, psychological challenges were much more complex and much tricker to deal with. We hope in the future, more research and enabling technologies can go beyond supporting day-to-day activities and focus on ensuring the mental well-being of workers under prolonged isolation.

Not everyone has the luxury of working from home, though. Many essential workers such as postmen, firefighters, doctors and policemen had to continue with their work on-site. And the service industry – restaurants, stores and other service providers – needs continuous human contact. To help these workers to protect themselves, a mobile app has been developed to detect if a person had been in close contact with an infected patient. This is done by leveraging constant anonymous Bluetooth data exchanges among smart phones. A user will receive a notification if they have been in cloud proximity (less than a few feet) for a prolonged period (more than 15 minutes) with an infected patient within the past two weeks.

When you travel during the pandemic, you are likely required to carry some sort of proof that you have not been infected. Some systems allow self-asserted health states, while some others require more formal proof such as a certificate issued by a doctor or proof of vaccination. This is a typical distributed trust problem – there are many issuers of proof with varied qualities, and there are many parties accepting different forms of proof. We are not sure where this will go at the end. One possibility is that proof issuers will be aggregated into a few trusted issuers. And these issuers can create a federation or consortium of trust so that their proof can be cross-validated. For instance, international travelers may be required to bear officially issued "vaccine passports" or even "vaccine visas" to travel to certain countries. Technologies like Blockchain and Distributed Digital Identity (DDID) are key enablers of such scenarios.

Searching for an effective vaccine and ways to cure the disease is an extremely complicated process with rigid scientific principles and massive computations. There have been some efforts of using crowdsourcing to understand how the virus works and to find possible cures. For example, protein folding is a compute-intensive process, which takes a sequence of amino acids and determine the 3D protein shape into which these acids fold. It's the key to understand how the virus work, and it requires a lot of computational power. Crowdsourcing projects split folding tasks into small chunks that can be shipped to personal computers around the world. Although we haven't found a cure yet, we remain faithful that we'll find effective vaccines, and we'll find ways to treat the infected.

1.4 SUMMARY

Edge computing is not "edge of the cloud". Instead, we define edge computing as "compute in context", which pushes computation to where the data is generated from the real world and how the result is applied. In other worlds, we believe edge computing is about fusing cyber world with the physical world.

Edge computing has a broad spectrum of application scenarios that have direct impact on making our lives better. This is because edge computing occurs on a massive, heterogeneous, distributed and dynamic compute plane that is intertwined with real-world contexts. This requires special considerations when you design edge computing solutions to ensure security, scalability and performance.

We'll discuss individual edge computing components in Chapter 2. Then, we will examine various design patterns in Chapters 3 and 4. In Chapter 5, we'll spend on Kubernetes due to its popularity. And finally, we'll discuss edge native design in Chapter 6 and how COA can help in Chapters 7 and 8.

Chapter 2

Edge computing factors

I am an Xbox fan. One of my friends is a PlayStation fan. You may have expected us to debate which console is better. However, we never do that. We both know very well the strength and the weakness of the platforms, so there's no point to debate that. We simply prefer one over another. My daughter, on the other hand, doesn't care about either Xbox or PlayStation – she likes her Nintendo Switch.

The different choices of gaming console reflect the reality of edge computing – edge computing is a very fragmented world. Or, if you want to use a positive word, a *heterogeneous* world. An edge computing stack requires all compute-related components, such as networking, storage, compute, software, security, policy management and many more. And for each of the components, there are multiple choices that come and go over time. This makes edge computing a dynamic, multidimensional space. To help you navigate this complex space, we'll take you through a journey from infrastructure management all the way to workload orchestration. Of course, we won't be able to provide a complete coverage. Chances are we'll probably miss something that you feel is utterly important. Regardless, we hope we can provide enough clarity on the landscape of edge computing and introduce you to the fundamental edge computing components, which we will use to assemble different design patterns in later chapters.

2.1 EDGE COMPUTING NETWORKING

Edge computing often involves multiple interconnected devices. Hence, networking is an important aspect of edge computing. Many research and infrastructural frameworks emphasize providing reliable and flexible connectivity among devices as well as between devices and cloud – this is where our journey starts.

2.1.1 Software-defined network (SDN)

Traditionally, network functionalities are provided by delegated networking devices (such as routers and switches) with integrated proprietary software.

This kind of hardware is hard to maintain, as you need to work directly with scattered hardware from different vendors using vendor-specific tools.

Software-defined network (SDN) separates *control plane* that manages the network and *data plane* through which the data flows. SDN enables network function virtualization (NFV) that replaces specialized hardware such as firewalls and load balancers with software components running on off-the-shelf server hardware (SDN nodes). SDN nodes can be centrally managed with standardized software stack. This brings some key benefits: end-to-end visibility and control from a centralized controller, flexible network topology that can be segmented and optimized for different workloads over the same physical connection, significantly reduced management cost and increased management agility and improved availability with redundant hardware and active failover to alternative connection.

An important application of SDN is SDN-WAN. Traditionally, reliable connections between a remote office and a datacenter are achieved by multi-protocol label switching (MPLS) circuits (often in a fixed hub-and-spoke topology), which are highly reliable but quite expensive. You also need to configure proprietary routers at your remote offices as well as in your data-center to maintain the connection. SD-WAN uses commodity server boxes (called SD-WAN nodes) to replace the routers. These boxes are managed by a central *controller* that publishes configurations and policies to these boxes to reconfigure the network over various physical connection types such as broadband and long-term evolution (LTE), as shown in Figure 2.1.

SDN, and SD-WAN overlay, underpins edge computing networks that connect on-premises servers, field gateways and server-grade edge devices to the cloud. Within a datacenter, *overlay* networks are often used to address some challenges in traditional datacenters, such as limited virtual local area network (VLAN) address space, lack of multi-tenancy support and bloating ToR MAC translation tables. Overlay network lays one or multiple virtualized networks on top of physical network to overcome these limitations.

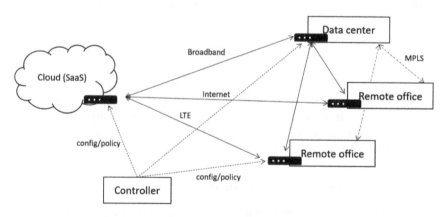

Figure 2.1 SD-WAN.

There have been various attempts of creating overlay networks, among which the VXLAN (virtual extensible LAN) protocol seems to have the broadest adoption. Another protocol is NVGRE (network virtualization using generic routing encapsulation), which is used by Azure virtual networks.

The core idea behind overlay networks is *tunneling* – network packets are captured, encapsulated and then forwarded to their actual destinations. A VXLAN package has an outer Internet Protocol (IP) header, a User Datagram Protocol (UDP) header, a VXLAN header and Ethernet header (header overhead is about 50 bytes per packet). The outer IP header is used by the underlying network infrastructure, and the destination Virtual Tunnel End Points uses the other headers to de-capsulate packets.

2.1.2 Sensor networks

A sensor network is comprised with many low-power, lightweight, small devices (called *nodes* or *sensor nodes* in this context) interconnected through wired or various types of wireless connections such as Wi-Fi, Bluetooth, cellular, radio and near field communication (NFC). A sensor network with wireless connections is also referred as a wireless sensor network (WSN).

WSN is a key technology that enables ambient intelligence. It enables scattered sensors to collect fine-granular environmental intelligence for scenarios like precision agriculture, healthcare, water quality monitoring, smart traffic control and many other scenarios. There are many possible topologies for a WSN, including peer-to-peer, ring, tree, linear, star and (partially connected or fully connected) mesh, as shown in Figure 2.2. Generally, the processing power of a node increases when the node is connected to more of other nodes on the network. For example, in a tree topology, a non-leaf node has more processing power than a leaf node.

On this network, a node that generates data is called a *source node*, while a data that receives data is called a *sink node*. A sink node can be

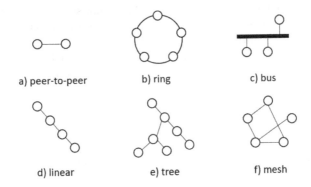

Figure 2.2 Sensor network topologies.

a sensor, a gateway or an aggregator that aggregates data from multiple sensors. A message often hops around multiple nodes before it reaches the sink such as a field gateway.

Sensor networks are often described by their application types. For example, a WSN used for transmitting personal data across personal devices in proximity is referred as a wireless personal area network (WPAN). Similarly, WSN for wearable devices is called a body area network (BAN). In both cases, sensors are often connected in a star topology as a user-owned mobile phone serving as an aggregator or a field gateway.

Sensor nodes on the network often need to operate on battery for extended period. So, it is important to reduce energy consumption as much as possible. Table 2.1 offers a side-by-side comparison of some low-power, low-bandwidth wireless network connection options.

Because of the power constraints, implementing property security is sometimes hard, because not all devices can afford running encryption algorithms such as 128-bit advanced encryption standard (AES)-based encryption. In such cases, an intrusion detection system can be used as a second-line defense that detects possible attacks by running rule-based or AI-powered anomaly detections.

2.1.3 5G

5G is the 5th-generation technology standard for broadband cellular networks. It offers low-latency and high-bandwidth connections that enable new edge computing scenarios, especially those need fast network speed with abundant bandwidth. 5G and its promise of high-bandwidth low-latency networking capabilities will dramatically change what is possible on the edge. Generally speaking, 5G is the next-generation wireless networking technology promising superior performance in nearly every category that is important for end user and edge scenarios. As shown in Table 2.2, the difference to LTE is quite remarkable.

We have already mentioned some of the scenarios that will be enabled through 5G in Chapter 1, but ITU-R's IMT-2020 standards provide a nice summary of the 5G service types and edge computing scenarios enabled by 5G networks, as summarized in Table 2.3.

5G uses high-frequency radio signals (high-band 5G uses frequency of 25–39 GHz, comparing to 4G that uses 600–700 MHz signals). The high

Table 2.1 Comparison of low-power communication protocols

Protocol	Range	Common data rates
Bluetooth LE	Less than 100 meters	125 kbit/s to 2 Mbit/s
Zigbee	10–100 meters	20 kbit/s to 250 kbit/s
Z-Wave	30–40 meters	Up to 100 kbit/s

Table 2.2 Comparison of 5G network and LTE network

	5G	LTE
Latency	I ms	30–50 ms
Throughput	10 Gbps/connection	100 Mbps/connection
Connections	1000000/km²	10000/km²
Mobility	500 km/h	350 km/h

frequency enables high data rate. However, the shorter wavelength signal has lower useful physical range and requires line of sight communication. This means the density of 5G towers is much higher than previous generations. Just for the record – around 2020, 5G's high-frequency signal actual stirred up some real fear in some communities, to the extent that some people went out and burned down 5G towers because they believed 5G signals would make them sick. Yep, we are at the dawn of auto driving and commercial space travel, and we still need to explain 5G cellphones and microwave ovens are different – one emits low-energy RF radiation (about 1 W) in the open and the other emits high-energy RF radiation (about 700 W) in a confined space.

The short working distance also means a device needs to switch towers more frequently when it roams around. More switches require more compute. And to make compute more efficient, improved chip technologies such as 7-nanometer chips or 5-nanometer chips are used to make 5G chips compact and efficient.

Traditionally, lots of engineering efforts have been spent to deal with low-bandwidth and instable connections. A pervasive, high-bandwidth

Table 2.3 5G scenarios

5G Service Types	Characteristics	Sample Scenarios
eMBB (Enhanced mobile broadband)	Stable connection Extreme high bandwidth (moderate rate for cell edge users) Moderate latency High spectral efficiency PER ~10^{-3}	AR/VR Ultra-HD 360° video gaming, conferencing
mMTC (Massive machine-type communication)	Extreme density (IoT), low power, low cost, low complexity, sporadically active, small data payloads PER ~10^{-1}	Sensor network Smart city Smart home Smart logistics Smart metering
URLLC (Ultra-high reliable and low latency)	Low latency, small data payloads, high reliability, high availability, strong security, often triggered by outside events such as alarms, bursty traffic PER ~10^{-5}	Self-driving car Industry automation Mission-critical apps Smart grid

network between cloud and edge not only enable scenarios as you've seen in Table 2.2, but also enables various hybrid scenarios that span cloud and edge. We'll discuss some of these scenarios in Chapters 3 and 4.

2.1.4 Long-distance networks

In many edge computing solutions, devices are deployed and used in remote places such as a water pump in rural areas or an offshore wind turbine. In such cases, running fiber cables to these devices is expensive and sometimes infeasible. Instead, low-power wide-area (LPWA) networks can be used to establish connectivity. LPWA uses long-wavelength (hence low-frequency) radio signals to carry data to great length. For example, the LoRa (short for long range) protocol uses sub-gigahertz radio frequency bands (433 MHz, 915 MHz, etc.) to send data to remote sensors that are kilometers away. Because LPWA connections have low bandwidth, transmitting a lot of data takes a long time and drains battery faster. So, LPWA is more suitable for scenarios in which sensors occasionally send small messages to the cloud (or base stations). There's also Random Phase Multiple Access (RPMA) technology, owned by a company named Ingenu. It uses the free 2.4 GHz bands to deliver about 50 times of bandwidth comparing to LPWA solutions, to the range about 50 kilometers.

To reach devices in extreme environments such as a weather station in high mountains or an oil rig in the ocean, we need a solution that can cover longer ranges. At the time of writing, multiple private companies are racing to launch Low Earth Orbit (LEO) satellites, with SpaceX leading the area with over 180 satellites (2020) in the sky. LEO satellites operate between 500 and 2,000 kilometers above the Earth. LEO satellites can be used as a centralized backhaul for 5G networks to cover a broad range of a geographic area.

That's about all we are going to say about networking. Next, we'll move up a layer and examine edge devices themselves.

2.2 EDGE DEVICES

Edge device is a very broad concept – essentially, any compute unit that can collect some data or perform some action in context can be called an edge device. Since this is a book focusing on software architecture, we won't spend much space on various device types. Instead, we'll focus on logical architecture of an edge device, starting with the physical device itself, which we refer as an *edge host*.

2.2.1 Edge host

In the context of this book, we care about compute unit that we can deploy workloads to. So, at the lower end of the spectrum of our discussion are microcontrollers, or microcontroller units (MCUs). MCUs are single-chip computers with their own central processing units (CPUs), memory and programmable general purpose I/O pins (GPIOs) that are often used for data collection, sensing and actuating. As the computer is often embedded in a bigger machinery, it's often referred as an embedded system. A MCU may also have a dedicated universal asynchronous receiver/transmitter (UART) for transmitting over serial line, a pulse-width modulation (PWM) block for driving analog circuits such as a motor and other communication ports such as universal serial bus (USB) and Ethernet. MCUs usually do not have an operation system. Instead, they simply run a single piece of code in a loop. Some more advanced MCUs run specialized real-time operation systems (RTOS) such as micro-controller operating systems (μC/OS, μC/OS-II and μC/OS-III), Mbed OS, FreeRTOS and others.

Moving a step up from MCUs are system on a chip (SOC), well, chips. These chips integrate more system functionalities on to a single chip – some of them have networking capabilities, some of them have a companion field programmable gate array (FPGA) that allows you to update logics on the chip, and some of them even have graphic adapters. SOCs are basically a computer on a single chip. It has obvious advantages such as smaller size, low cost and moderate energy consumption. And the disadvantages of SOC include complexity, rigid configuration and limited resources. Move up a step further, we have system on a module (SOM), which is a board with several chips that deliver compute functionalities. A SOM board is often paired with a carrier board (with Ethernet ports, USB ports, power and display ports, etc.) to form a complete computer. Then, we have single board computers (SBC) that are computers on a single circuit board. There are also other packaging form factors, which we won't enumerate here. From software's perspective, we uniformly view them as computing units with various capabilities. This view will become clearer in Chapters 7 and 8 when we discuss capability-oriented architecture (COA).

Keep moving up the ladder, you'll see more familiar compute devices that run various Linux distributions (including custom-built Yocto distributions) or Windows OS. And as you move into PC-level devices, you have a mixture of consumer devices such as laptops, tablets, netbooks, desktop towers, plus various mobile devices and wearable devices such as smart watches. Also in the mix are various types of accessories and specialized computing devices such as 3D printers and VR/AR goggles.

As you continue "zooming out", you see server-grade edge devices. These are powerful edge servers that often serve as field gateways, aggregation points and on-site control centers. And eventually, you can assemble these devices into clusters to provide an available and reliable computing system.

2.2.2 Virtualization

Although not mandatory, there is often a virtualization layer between the host and the workload packages. For example, Project EVE from

Linux foundation allows you to run multiple operating systems as virtual machines on the same hardware (it also supports containers, which we will discuss further in Chapter 5).

The first benefit of virtualization is consistency, because you can capture your entire software stack, including the OS and all required libraries and desired configurations, as a system image that can be consistently applied to multiple devices. This kind of consistency is critical for device management at scale. The system image can also be digitally signed and verified before launched. This ensures that system has not been tempered when it's launched.

Throughout computer science history, people have been trying different ways to reinforce consistency across multiple devices with technologies such as installation scripts, infrastructure as code solutions, auto system reset (that re-image the system upon reboot, such as a Cybernet Ghost Drive), virtualization and containerization.

The second benefit of virtualization is isolation. When you have multiple applications share the same set of hardware resources, you want to constraint how much an application can consume so that it does not hog all resources, also known as *noisy neighbor* problem. You also want to isolate the applications for security and privacy reasons.

At the time of writing, there are two major flavors of hypervisors, *type 1* hypervisor and *type 2* hypervisor. Type 1 hypervisor runs on bare metal, and type 2 hypervisor runs on top of an operation system. There are also hypervisors that are specially designed to support running containers. These containers have their dedicated system kernels (comparing to containers with shared kernels) so they are better isolated from each other.

2.2.3 Software stack

When you boot up your device, the system's *firmware* (such as basic input/output system – BIOS and the newer unified extensible firmware interface – UEFI) is loaded and executed. The firmware verifies and initializes hardware, and it goes through storage devices (and network interfaces) to search for a *bootloader* (such as the GRUB bootloader for Linux). The bootloader in turn loads the operation system (such as Linux, RTOS and Microsoft Windows).

When it comes to operation systems, there are two major flavors: RTOS and "regular" operation systems (OS). The key difference between an RTOS and an OS is that RTOS guarantees predictive response time without jitters. For example, when you schedule a task to be triggered every second, RTOS guarantees that task is triggered at every second (with exceedingly small variances), while a regular OS, with the exception of real-time extensions for Linux OS, doesn't give you such guarantee – when the CPU is busy, a scheduled task might be delayed.

One of the most popular RTOS systems is FreeRTOS, which is used by AWS IoT devices such as AWS IoT Core and AWS IoT Greengrass. Azure, on the other hand, uses Azure RTOS ThreadX.

The Internet of Things (IoT) applications are often written in C and assembly language, which generates smaller binaries that can be fit onto limited resources of edge devices. Those low-level languages also do not rely on a garbage collected runtime, which causes jitters that is undesirable in near real-time applications. More powerful devices often support high-level programming languages to make programming edge devices more approachable. Java virtual machine (JVM) has been a popular runtime choice. And Python has gained much attention in recent years because of AI applications. Microsoft. NET Core has gained some tractions as an alternative runtime as well.

There are often several agents running on an IoT device. These agents are injected by different systems to make connections to backend management systems and data pipelines through various protocols such as Hypertext Transfer Protocol (HTTP), Remote Procedure Calls (gPRC) and Message Queuing Telemetry Transport (MQTT). Some other agents are responsible for communicating with local devices and accessories – such as collecting data from sensors. Agents on the same device don't often communicate with each other. When they do, they usually communicate through a local message bus, as shown in Figure 2.3.

Just like on any other computer systems, software on edge devices also needs to be updated. Updating a device usually means to re-image the whole system, to apply a new container or to run an update script. To avoid a device becoming "bricked" (which means the device is corrupted in a way that can't be restored to a functional state), a reliable update process uses an A/B upgrade scheme, which installs the new version side-by-side with the original version and then switches to the new version. If the new version doesn't work as expected, the system is rolled back to the original version. For example, when re-imaging a system, the new system image is written to a separate disk partition. Then, the bootloader is marked to load from

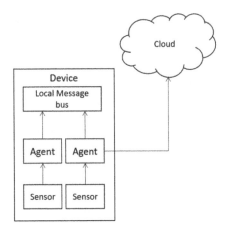

Figure 2.3 Asynchronous agent communication pattern.

the new partition and the device is rebooted. When the device is rebooted, if the new system doesn't load correctly, the bootloader switches back to the original partition.

The software stack on a device can be tampered. Especially, because many devices are installed in open environments such as streets and shared spaces, they are subject to physical attacks such as modifying the content on storage cards, altering the circuit or installing interfering devices. We'll discuss device security in the next section.

2.2.4 Device security

When devices are installed on the field, they are completely exposed to attackers. Because the entire stack may be altered in various ways, nothing can be trusted without proper verification. Of course, at the end you still need to trust something, otherwise no verifications are possible. However, you want to constrain that thing you trust to be extremely resilient to all possible attacks. Then, you can use this thing as the root of your trust chain that you can use to verify the entire hardware and software stack.

Trusted platform modules (TPMs) are secured crypto-processors that can be used to provide system integration checks using integrated cryptographic keys. Nowadays, most modern laptops are shipped with TPMs that can be used as root of trust. TPMs can be used to verify if the firmware and the operation system have been loaded properly without modifications. A TPM is manufactured with a unique *endorsement key* (EK), which is a public/private key pair burned onto the device. The EK is signed by a trusted Certification Authority (CA). At load-time, TPM calculates a hash of executables and other files to "measure" these artifacts. Then, the measurement (along with TPM public key) can be sent to a remote attestation service to verify if data integrity has not been broken (by calculating a hash of certified artifact sets and comparing with the hash code it receives, for instance).

Device authentication often uses public/private key pairs such as a X.509 certificate. To securely save the private key on device, you can use a hardware security module (HSM), which is a piece of temper-resistant hardware that can be used to store secrets and to perform encryption/decryption and digital signature operations. When a device is manufactured, it is often provisioned with a bootstrapping certificate. Upon the first boot, the device contacts a predefined rendezvous endpoint and uses the bootstrap certificate to authenticate and exchange for an operational certificate that is used for later operations. For example, the operation certificate can be used to authenticate with various cloud services the device uses.

Other than secret keys, there are also other valuable on-device assets that need protection: user data, machine learning models and other intellectual properties (IPs) contained in the device binary or source code. The first thing you need to ensure is that the code or the machine learning models you run on the device have not been altered. This can be done by verifying signatures of the corresponding software package. You may also want to obfuscate text-based code to make extracting your IP a bit harder. For in-depth protection at runtime, you may want to consider running critical code in a trusted execution environment (TEE), which is isolated from the main process and offers a secured enclave to run code. Finally, to protect user data, you need to consider from three vectors: at rest, in transition and in use. Generally, you want to save sensitive data encrypted on disk, sometimes with customer-controlled keys. And when you transmit data across any boundaries, you need to ensure data is exchanged through secured channels. And finally, to protect data in use, you may want to consider TEE, or employ cryptographic algorithms that make it difficult to sniff out secrets by analyzing data access patterns or time delays.

Instead of being completely broken, a device may malfunction due to various reasons such as hardware/software failures and hacker attacks. Hence, it is often useful to be able to monitor devices and detect abnormalities using analytical, statistical or machine learning methods. When a device is malfunctioning, you may need means to directly attach to the device to diagnose problems. In such cases, you need to ensure there's proper user authentication protocol in place (and please do not use a default password) to make sure only authenticated users can access the device. You may also want to implement just-in-time (JIT) access policies so that a user gains minimum required access only during the troubleshooting process.

2.2.5 Intelligent devices

I used to work in a small office in California. The office was equipped with motion sensors that got automatically activated after office hours. Once, as I was wrapping up work late in the office, I realized the motion sensor by the front door had been activated, and my trying to exit would have triggered the security company with whom we have a first-respond contract. Fortunately, I knew the sensor was a microwave sensor, which detected motion by measuring the reflect changes in a high-frequency signal it emitted. So, I stood up and moved very slowly toward the door to avoid sudden reflection spikes. I was able to exit the office with no problems that night. If the company had used a passive infrared (PIR) sensor that detected body heat, or a camera that detected motion in visual frames, I would have more trouble tricking the system. That was over 10 years ago. Although the system was not perfect, it demonstrated how people had been seeking intelligent edge solutions at different scales.

Many of the most exciting edge computing projects have an aspect of intelligence on edge. And intelligence on edge often has several unique constraints:

- Power consumption. To reduce the maintenance costs, edge devices are often required to run on batteries for an extended period of time. Running complex algorithms consumes more energy and sending more data from the devices to cloud incurs more costs.

- Response time. Edge computing is where the cyber world collides with the physical world. Some scenarios such as self-driving cars and guided missiles require very tight response loops. This makes it infeasible to offload work to a remote device or server.
- Privacy concerns. People don't usually like being watched or listened to, especially when they are in private environments such as their homes. And no business likes their business secrets being accidentally leaked by their surveillance systems. Hence, privacy protection is often a required feature of any intelligent system.
- Form factor. Devices need to be adapted to the environments where they are installed. They may need to endure a lower temperature, high pressure, water, sand and other hostile environments. Furthermore, they need to be fit into the bigger machineries they serve. All these heavily impact how the devices are designed and manufactured.
- Cost. Besides everything else, device makers produce devices to make a profit. They carefully review their bill of materials to cut any waste. Progress in material and manufacturing technologies have drastically reduced common costs of intelligent devices in the past decades. However, as people seek more sophisticated scenarios, the struggle between controlling cost and building capabilities continues.

To satisfy the latency requirements, different hardware have been developed over the years, with graphics processing unit (GPU) playing a significant role. GPU was original designed to accelerate compute graphics for simulation, computer-aided design (CAD) and gaming scenarios. CPU processes one instruction on a single piece of data at a time. However, GPU can apply a single instruction on a collection of data at the same time (which is called SIMD – single instruction, multiple data, which is also seen in modern CPUs). This makes GPU suitable to do parallel processing on large amounts of data, which is required by machine learning training and inferences. In recent years, multiple neural processing unit (NPU) processors have been designed to accelerate AI operations, such as the tensor processing unit (TPU) from Google, Ascend from Huawei, Inferentia from AWS, Graphcore intelligence processing unit (IPU) from Azure and many others. These processors are application-specific integrated circuits (ASICs), which are designed and optimized for specific tasks and can't be changed after manufacture. FPGAs, on the other hand, allow its processing units to be reconfigured by a customer after manufacture. They offer great flexibility to adapt to different algorithms but not as optimized as ASICs.

Machine learning models are often built and trained with floating-point numbers. To reduce the size of the models and to make inferences more efficient, you can use a process called quantization to transform the AI models to use low-bit integers. Although you usually lose some precision or recall during the process, the reduction is often significant in size and power consumption. There are also toolkits like OpenVINO toolkit that can optimize machine learning for specific hardware. OpenVINO supports popular frameworks such as TensorFlow, Caffe and ONNX and several hardware platforms such as Intel Movidius VPU.

Before we wrap up this section, we'd like to mention a new generation of photonic processors that use light to process and transport data. Essentially, data is encoded as "brightness" of laser beams passing through the circuit. Such circuits consume less energy and offer better performance. And what's most important is that calculating with light is pretty cool.

Next, we'll zoom out once again and examine how devices are managed at scale.

2.3 FLEET MANAGEMENT

On November 7, 2020, US President-elect Joe Biden celebrated his projected victory of winning the election with fireworks and a drone light show. The drone light show used a few hundred coordinated drones to form figures and letters in sky. The bright, dynamic 3D figures in the dark sky brought smiles to faces and somehow a faint hope for the better into people's hearts. Ever since its first appearance in 2012, drone light show has gone brighter and bigger. At the time of writing, China holds the Guinness World Record of a drone light show with 3,051 coordinated drones.

Coordinated drones demonstrate the power of a device fleet. When multiple devices are orchestrated, they can achieve amazing scenarios such as coordinated search and rescue, smart stadiums and intelligent cities. This section surveys technologies that coordinate actions of multiple devices.

2.3.1 Clustering

A computer cluster is comprised with more than one compute unites. This compute unites assume two different roles: *manager role* and *worker role*. The manager role is responsible for monitoring and maintaining the overall health of the cluster. And the worker role is responsible for carrying out the actual work. There's usually a single *primary* manager and multiple *secondary* managers as backups. The manager runs a *job scheduler* that schedules tasks to workers. Worker roles are usually homogeneous, which makes work scheduling simpler. Some more complex clusters may contain multiple classes of worker roles with different capacities. And some sophisticated scheduler support attribute-aware scheduling that can schedule workloads based on properties associated with workers and workloads. For example, a workload may tag itself as requiring GPU. Then, the scheduler will attempt to place the workload on a GPU-enabled worker. Similarly, workloads can use attributes to indicate they should be co-located (such as a data analysis service and a database) or exclusively placed (such as two CPU-intensive tasks).

Sometimes, it's helpful to separate sensors that collect data and workers that process data. This is a key design to simplify data collection units and

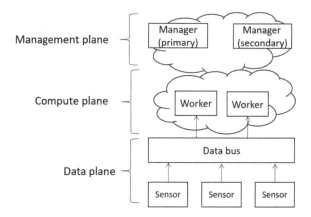

Figure 2.4 A cluster with three planes.

achieve high-available, flexible data processing functionalities. Figure 2.4 illustrates such as cluster. In this cluster, data collection sensors and actuators don't have any sophisticated logics on them – all they do is to collect data or to react to commands. Collected data is sent to a processing cluster through a data bus. The workers on the processing cluster pick up data messages from the bus and process them. This allows all processors to be centrally managed. And multiple processors can be used to both scaling out processing power and serving as backups to each other. Informally, we can segment the cluster into three planes: a management plane on which cluster managing commands flow, a compute plan on which workload traffic flows and a data plane on which sensor data flows.

Modern cars contain a number of electronic control units (ECUs) that manage electrical systems or subsystems in a vehicle, such as a brake control module (BCM), a suspension control module (SCM), a transmission control module (TCM) and a few others. The architecture shown in Figure 2.4 allows to break out compute tasks that don't require automotive-specific safety, security and hard real-time capabilities on these ECUs to be centralized into a centralized compute plane that can use commodity hardware that forms a cluster (such as a Kubernetes cluster, the topic for Chapter 5) to run applications written in high-level language. This design has many benefits, including (but not limited to) higher availability, lower complexity, richer and customizable user experience, easier software development and maintenance (such as applying software updates).

2.3.2 Identities and access control

When a device is managed by a system, it needs to establish its identity with the system. When we discussed device security earlier, we've mentioned that a device can use a certificate imprinted by the builder as its identity, or as a proof to acquire a proper identity from the management system. However, when the device is manufactured, its final owner is still yet to be defined (as it's not sold to a customer yet). This is where a provisioning service comes in. A provision service provides a well-known URL that a device can invoke when it is first turned on. And the provision service itself can be configured to connect to the destination management systems. Essentially, the provisioning service serves as a broker to onboard the device onto appropriate systems.

Things become more complicated when a device is handed from one system to another. Usually, the device needs to go through another onboarding process to get registered with the new system. This is when a device gets a new identity and continues with its life cycle in the new system. However, in most cases, such handoff is not a clean cut. For example, the original owner of the device may have maintenance obligations that need to be tracked. In this case, the device has multiple identities – one for the original owner to track maintenance, and one for the new owner to manage continuous operation.

In connected scenarios, a device often needs to communicate with one or multiple cloud services, which require different types of authentication and authorization, such as digital certificates, shared keys and service credentials. Some of these credentials are time-bound, which means they must be dynamically acquired or periodically refreshed. This requires the device to hold on to some piece of secret so that it can use the evidence to exchange for access keys. As we've mentioned earlier, the device can save the secret in secured hardware like HSM. And the key exchange can happen over traditional secured channels, or through quantum algorithms such as BB84 and E91.

In traditional service-oriented systems, role-based access control (RBAC) is often used as a way to manage user accesses at scale. Instead of creating explicit access control lists (ACLs) for all users, users are grouped into roles, to which certain access privileges are assigned. In the context of edge computing, a more appropriate access control method is attribute-based access control (ABAC). With ABAC, accesses are granted or denied based on the attribute values a device can present. ABAC works better than RBAC because RBAC still relies on administrators to manually configure role accesses and role memberships. ABAC, on the other hand, allows access control over the current states of devices. For example, to revoke access to a service from a device with a lower software version, RBAC requires moving the device out of the corresponding role. On the other hand, ABAC allows rejecting access based on the device's software version without human intervention.

Authentication identifies a device. Authorization devices what a device can do. However, these are not enough to control how a device behaves. Next, we'll examine how to reinforce consistent configurations and policies across a fleet of devices.

2.3.3 Configurations and policies

Configurations and policies are centrally managed and pushed to devices through a distribution system. Because most devices can only make outbound connections, they often need to periodically query the central management system to get updated configurations. The new configuration is applied through a *state reconciliation* process: a device compares its current state with the desired state retrieved from the server and performs necessary actions to drive its current state toward the desired state.

Because each device drives its own state reconciliation process, a configuration is applied to all devices in an *eventually consistent* manner. During the update process, some devices will have updated configuration while some devices will have the original configuration. This may cause confusion, especially the new configuration drives the device to use a different message schema. Hence, it is a good practice to associate versions to all the metadata and data you pass across the system, and to ensure backward compatibility in your service implementation. Because some devices have long deployment cycles and long life spans, you may need to ensure backward compatibility with many older revisions. On the other hand, you also want to employ anti-rollback protection that prevents attackers from loading older software that has known vulnerabilities.

Policies are often associated with compliance requirements to make sure the devices stay compliant to corporate and industry policies. The difference between policies and configurations is that policy items are proactively checked by the system, while configuration items are not repeatedly checked by the management system.

One thing to keep in mind is that configuration is also an integral part of device software stack that affects how device behaves. You should treat configuration update as serious as a software update. In theory, software update and configuration update should be performed by the same mechanism. Unfortunately, updates at different levels are often carried out by different systems: firmware and OS updates are performed by re-imaging the device; software libraries are updated through install scripts and containers are updated through obtaining newer images from a container registry. Later in this chapter, we'll present some ideas to create certain uniformity across these processes.

Now, since we've examined different layers of edge computing in a static view, we'll shift the focus to examine the complete device life cycle through time.

2.4 PERSPECTIVES AND UNIFORMITY

Throughout the device life cycle, different people come from different perspectives focusing on different things. Before we go further, let's meet these people.

2.4.1 Human players

Just like many other activities in modern society, device making and operation is a collaborative process that involves several different roles:

- **Silicon vendors:** Silicon vendors are the person who makes integrated chips. They supply the chips to device makers.
- **Device makers:** Device makers make the physical devices.
- **ISVs:** ISVs and other software vendors supply software to run on devices, including operation systems, frameworks and application packages.
- **System integrators:** They implement edge solutions for customers. They may purchase devices from device makers and then install, configure and sometimes operate the devices on the customer's behalf.
- **Operators:** Operators keep the devices operational and perform day-to-day monitoring and management tasks.
- **Data scientists:** Data scientists are playing significant roles in edge computing. They extract value from the data stream, and design intelligent models to be achieved intelligent edge computing scenarios.
- **Customers:** Customers purchase and use the devices.

The role distinction is not strict. For example, many customers operate their own devices, so they are assuming the operator role as well. And we are sure we are missing some other people who also play significant roles during the device life cycle. Regardless, the key message here is not to enumerate all human roles, but to illustrate the fact that there are multiple roles coming in from different perspectives to operate on the same physical device.

2.4.2 Multiple perspectives

Data scientists rarely care about how the operation system is designed. And a chip designer does not care about Kubernetes, regardless how popular it is in the microservices community. Separation of labor allows bigger achievements in individual areas. So, different people coming from different perspective are a healthy thing. They bring their expertise to generate more value in the process. Moreover, people often come up with different solutions to the same problem. And it's very unlikely that a single solution will take over the entire market share. The norm of edge computing is that this is, and will always be, a segmented, heterogeneous and dynamic landscape. However, as a single device going through the life cycle and being operated by different people, it's very desirable to create some sort of continuum so that we can ensure accumulation of value instead of different efforts canceling or forbidding each other – such as a function-rich library being too big to fit into smaller devices, or an inefficient chip design forbidding high-performant machine learning models.

As the last topic of the chapter, we'll explore the idea of creating some sort of uniformity out of the multiple perspectives.

2.4.3 Uniformity

Before we talk about creating uniformity, we need to first answer the question of if we need to create uniformity. If the edge compute world will always be segmented, are there any points to even attempt doing so? Our vision here is certainly not to unify the entire edge computing world. Our goal is to create uniformity in a smaller scope, such as a software-as-service (SaaS) solution or a device product line. For example, we want to help answer the questions like: if I made a SaaS service, how can I make sure it works with a broad range of devices with a consistent customer experience?

Once we clear that up, we need to find the stem around which we'll build the uniformity. Although it might work for small projects, unification on a single technical stack rarely works well in a moderately complex edge computing solution. Instead, we believe the thing we can use as the stem is *workflow*. For example, when you work on a machine learning solution, you may use various data collection and tagging technologies, different ways to design and train your model using different AI frameworks on top of different hardware, and deploy your trained model using different means – such as a web service or a container. However, the general workflow is the same: you collect data, train your model, publish the model, monitor it, re-train it and publish it again. The workflow we just described is agnostic to any platforms, frameworks or hardware. It represents a generalization of how a machine learning project progresses. And this is how we can build up uniformity. The workflow is comprised by a number of *steps*. And each step requires certain *capabilities* to carry out the required tasks. Capabilities are in turn delivered by *vendors* who provide solid implementations on top of actual hardware or software stack.

Another key to create uniformity across different aspects is to create a unified language. We certainly don't mean a unified messaging format or communication protocol. Instead, we believe what's missing is a semantic description of capabilities that can be understood by all participants in the ecosystem. This semantic description works like English (or any natural language). It carries the semantic meaning of entities and actions – like "dogs" or "eat". Skeptics have told us that such description has too much ambiguity and is not sufficient to support the actual information exchange across systems. However, they are neglecting the fact that this language is not designed to facilitate data transmission. Instead, it's aimed at creating commonly understood semantic components that can be used to describe generic workflows. We'll explore these ideas further in Chapters 7 and 8.

2.5 SUMMARY

This chapter offers a brief survey of various factors of edge computing, from networking to device to fleet management. For networking, SDN plays an important role in edge computing. And a mixture of different networking technologies and protocols enables the pervasive connectivity required by a ubiquitous edge computing plane. For devices, the chapter walks up the layers from hardware to virtualization to upper-level software stack. Then, it introduces how security and intelligent edge is applied. Finally, the chapter reviews different human players and the roles they play. Then, it suggests some initial thoughts on how to create uniformity out of the chaotic edge computing landscape.

COA, which is the main topic of Chapters 7 and 8, will further explore these interesting ideas and propose a framework to build a unified edge computing platform.

Section II

Edge computing design patterns

Chapter 3

Edge to cloud

As the dust settled, the crew of the Invincible woke up from their deep sleep and gazed through the fogged portholes at the barren landscape of Regis III. Their mission was to investigate the loss of Condor, a sister ship, on this seemingly lifeless planet. To their surprise, they soon discovered that the planet was inhabited by insect-like tiny machines that moved in swarms. When they were in small groups, they were pretty much harmless. However, when they felt threatened, they could form huge clouds and overwhelm their opponents with swift attacks and huge surges of electromagnetic interference that apparently wiped out not only the electronics but also memories.

As the Invincible crew stepped into their unknown fate in Stanisław Lem's 1964 science fiction, the Invincible, the situation on planet Earth in a parallel universe was far less devastating. Here on Earth, people have been fascinated by an idea of *Smartdust*, which is comprised with tiny microelectromechanical systems (MEMS) that can be scattered in the real world without being noticed. They can remain dormant for months or even years and be activated by an external energy source (such as a probing laser from a flyby airplane). The amount of the energy carried by the laser beam is just enough to wake them up, collect some environmental data and send the data back to the prober. They can be used to plot accurate local maps of chemical or radioactive contaminations, temperature patterns, soil moisture and PH values, ground vibrations and many other environmental information.

Smartdust represents the philosophy of the initial wave of edge computing – to extract value from connected data sentinels. Many Internet of Things (IoT) projects are centered around this idea. In such projects, devices are used as collectors of data, which are then fed into a data pipeline that aggregates data from multiple devices and performs various analyses to extract values, such as pattern and anomaly detection, trend analysis, prediction and optimization. Because this pattern is so dominant in edge computing, some believe "edge" simply means "edge of cloud". As we explained in Chapter 1, this is a rather limited view of edge computing. However, it does indicate that a close collaboration between cloud and edge plays an important role in many edge computing solutions.

In this chapter, we'll examine several common patterns of edge-cloud collaborations. Especially, we'll focus on how data and compute are concentrated toward the cloud in these patterns. In the next chapter, we'll examine the other direction – how compute and data are dispersed into edge.

3.1 DATA COLLECTION PATTERNS

The main goal of this pattern is to facilitate data collection from edge devices to cloud. Although there are often device management, configuration management and command & control aspects associated with these solutions, this pattern differs from others because it's not focused to push compute to edge but to get data back to cloud.

Figure 3.1 illustrates a high-level architecture of the data collection pattern. On the cloud side, a data ingress endpoint is configured to accept data from devices. This ingress endpoint is invoked by either direct invocations or a messaging bus. Devices are connected to the ingress in several different ways. *Device A* is directly connected to the ingress endpoint; *Device B* and *Device C* are connected through a field gateway (or a base station), which aggregates data from multiple sensors and provides additional capabilities such as batching, ordering and filtering. Finally, *Device D*'s connection is bridged by a local sensor network such as Zigbee, wireless sensor network, Narrowband Internet of Things (NB-IoT) and LoRaWAN before it reaches either the field gateway or the cloud.

Next, we'll examine some of the popular data collection products and techniques and examine how they establish and maintain scalable and reliable data pipelines to collect data from edge to cloud.

3.1.1 Data collection through messaging

Collecting device data through messaging is a popular choice. Systems like Azure IoT Edge, AWS IoT Core and Apache Kafka all support collecting device data as messages. And there are widely adopted messaging protocols

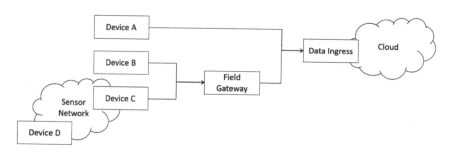

Figure 3.1 Data collection pattern.

such as Advanced Message Queueing Protocol (AMQP) and Message Queuing Telemetry Transport (MQTT).

In such systems, a *publish-subscribe* pattern is often used to decouple data sources and data consumers. Data sources *publish* messages to *topics*. And data consumers *subscribe* to topics that they are interested in. Messages are queued on topics for consumers to pick up. Messages are appended to the end of topic queues, creating hot write zones on disks. For better scalability, topics can be segmented into multiple *partitions*. This allows more efficient parallel writes to the same topic. However, the price is that ordering of messages across partitions is often not guaranteed in such cases.

On the consumer side, a consumer maintains a persistent marker that points to the last message it has processed. The marker allows a consumer to pick where it has left off – when it crashed, for example. Further, multiple consumers can be organized into a *consumer group*. Consumers in a consumer group read from the same topic partition. This allows a partition to be drained quicker. The consumers in the same group also serve as backups of each other. When one consumer fails, others can keep the message pipe going.

Most messaging systems ensure *at-least-once* delivery, which means any message is guaranteed to be delivered to a consumer at least once (but could also be delivered multiple times). In some cases, the delivery guarantee can be relaxed to *best-effort*, in which messages can be discarded after a few failed delivery attempts. In systems that require near real-time reactions to temporal data, using a best-effort policy can help the system from becoming too bogged down by a few malformed messages or spotty networks. Generally speaking, achieving *exactly-once* delivery is very difficult in a fault-tolerant system. Hence consumers need to be designed to handle duplicated messages gracefully, such as by implementing *stateful idempotent functions* that can be invoked with the same input many times and always generate the same output. For completeness, there's also an *at-most-once* delivery, in which a message may or may not be delivered but never duplicated. Think of sending a package to your friend – the package may never arrive, and it will never be duplicated.

Messages are sent to topics either in *batches* or as a *stream* (or batch-of-1) of messages. Batches help to improve *throughput* and streams helps to reduce *latency*. Based on your scenario, you may choose between the two modes and adjust batch sizes to achieve the best result.

The messaging system decouples message sender and consumers, in the sense that senders and consumers don't directly invoke each other. They can be scaled and hosted independently from each other. And replacing a sender or consumer (even dynamically) doesn't affect the other party. However, message senders and consumers are still coupled by data schemas, as consumers need to understand what the senders are sending to make sense of the data. Many messaging systems have *envelops* or *metadata* associated with message payloads so that consumers can infer important information about the message payload – such as schema version – before attempting to process the message.

When a messaging system is used, it's required to be secured, reliable and scalable. Maintaining a production-grade messaging system is not an easy task. Cloud-based message systems hide most of the operational complexity from you and offer highly scalable and reliable messaging buses for your solutions. Some messaging systems are also integrated with device management systems so that devices can authenticate with the backend system with provisioned device identities. All these characteristics make cloud-based messaging systems a preferred choice of IoT scenarios.

3.1.2 Data collection through ingress endpoints

Data collection can also be performed through direct service invocations to a data ingress endpoint through common protocols such as Hypertext Transfer Protocol (HTTP) and File Transfer Protocol (FTP). This approach is most suitable for situations in which you need to upload large data files that exceed the size limit of common messaging systems.

Dedicated ingress data points are also used for streaming scenarios, such as streaming surveillance camera feeds to cloud through media streaming protocols such as Real-Time Streaming Protocol (RTSP), Real-Time Messaging Protocol (RTMP), HTTP Live Streaming (HLS) and MPEG-DASH.

Using a messaging system often needs a client library that supports the specific protocol or backend. Such libraries are not always available on all devices. On the other hand, for internet-enabled devices, making Transmission Control Protocol (TCP) or HTTP connection is a fundamental capability. Using an ingress endpoint with REST API lowers the bar of a device connecting to the backend. A REST API is efficient for scenarios of many devices making occasional connections to the backend.

An ingress endpoint can be scaled out by simple load balancing or partitioning. In the case of simple load balancing, the ingress endpoint is backed by multiple service instances behind a load balancer, and client requests are distributed evenly to backend instances. In the case of partitioning, service instances are partitioned by certain key values and matching data elements are routed to corresponding partitions.

An ingress service is often stateless, which means it doesn't save any local states. To ensure availability of an ingress endpoint, multiple ingress service instances are used as backups of each other. Once the data is received, a service instance often saves it to a high-available data store, such as a replicated key-value store, for persistence. Please see the next section for more details on how data continues with a server-side pipeline once it's ingested.

Ingress endpoints are sometimes used in conjunction with a messaging system. In such a configuration, the ingress endpoint provides a simpler REST API for the devices. It also abstracts the messaging platform and messaging protocol from the devices.

3.1.3 Bulk data transportation and in-context data processing

The bulk data transportation pattern is originally used to allow customers to transfer large amount (terabytes or petabytes) of data through a specialized data transfer device that can be physically shipped between the customers and the cloud. Later, the pattern evolved to allow certain compute tasks to be carried out on these devices locally to provide fast, private, in-context data processing.

Amazon Web Services (AWS) Snowball devices, including Snowmobile, Snowball and Snowball Edge, are pre-built, secured hardware with storage and compute capacities. They are shipped to customer sites to run on customers' local network. And as they are filled up with data, they are shipped back to AWS and the data is made available on cloud. Snowball devices are not designed for continuous data collection. Instead, their main purpose is to migrate a large amount of data to cloud. Later, Snowball Edge devices support running local EC2 instances and Lambda functions. You can join multiple Snowball Edge devices into a cluster for more complex workloads or higher storage capacity.

Similarly, Azure Data Box device family – Data Box, Data Box Disk and Data Box Heavy, is designed to facilitate data transfer between on-premises and cloud. Azure Data Box Edge (which is later rebranded to Azure Stack Edge) adds compute capabilities by including on-board field-programmable gate array (FPGA) and graphics processing unit (GPU) on the device to run on-device machine learning and data processing operations.

The evolution from bulk data transportation solutions to edge computing solutions is quite interesting. If data can be processed in context and then later consumed in context within a customer's environment there really isn't much value to make the round trip to cloud, except for the purpose of long-term archiving. And this is precisely why we don't believe edge computing is all about adding an edge to the cloud. Edge computing is about processing in context. And cloud may or may not be in this process.

3.1.4 Data pipeline on cloud

Once data reaches cloud, it goes through a data pipeline that extracts value from the data stream. A data pipeline is comprised of the following components (not necessarily in the same order):

- Data transformation

Data transformation focuses on transforming data into a format that can be handled by a downstream processing unit. When a system ingests data from multiple data sources, data packets from different sources often need to be *normalized* into a generic format so that they can be aggregated and analyzed in a consistent manner.

In addition to normalization, other common data transformation patterns include *augmentation*, in which data fields from a secondary data source is used to augment records in the primary data stream; *translation*, in which data packets are translated from one schema to another schema based on a *projection rule*; *correlation*, in which data packets from different sources are assembled into combined records based on certain correlation keys; and *projection*, in which data values are projected from one value domain to another value domain.

- Data processing

Data processing is a broad category that covers all sorts of computations applied to the data stream. At the time of writing, Spark is one of the most actively developed and used distributed computing engines. It provides a unified computing engine that supports a wide range of data analysis tasks, ranging from simple Structured Query Language (SQL) queries to complex AI and streaming computations. Before Spark came along in around 2014, Hadoop was the uninterrupted leader in big data processing. Its MapReduce API, as well as various programming models built on top of it (such as Pig, Hive, Cascading and Crunch) was widely used in big data projects.

- Data storage

Data is often persisted for future uses. Although relational databases are still used, more projects elect to use NoSQL databases instead. NoSQL databases often use an eventual consistency model, which allows them to provide shorter response times and higher throughputs compared to traditional relational databases. NoSQL databases also support flexible schemas, so they are more tolerant to data format changes. Some projects use *column store databases*, which organize data into column families instead of rows. Column store databases are ideal for data compression, aggregation queries and massive parallel processing, making them popular choices in big data scenarios. Furthermore, a time series database (TSDB) is optimized for storing time series data. This makes it ideal for handling timestamped sensor readings.

Another axis to examine the data storage options is by the "hotness" of data. *Hot* data is the most frequently accessed data. This type of data is often kept in memory or solid-state drives (SSDs) for fast read/write accesses. *Warm* data is kept in traditional hard disk drives (HDDs) for analysis and queries. And finally, *cold* data is kept on tapes or newer cold storage media such as silicon (that stores data as etched vectors or holograms) and DNA proteins (that encodes data as DNA sequences).

- Data visualization

The goal of data visualization is to present raw data and analysis results to human users so that they can take further actions based on the intelligence.

Live data stream views allow user to monitor the current status of devices; *pre-populated views* present aggregated information or analysis results; *query tools* allow user to run dynamic queries using T-SQL queries, property filters (to filter key-value pairs) and graph-based queries. In addition, *triggers* can be defined to automatically trigger actions when certain threshold is met.

Excessive amount of data will overwhelm human users. Nowadays, more and more systems try to apply certain machine learning techniques to automate data processing with minimum human intervention. For example, a monitor system may automatically triage system alerts and run automated actions based on predefined policies. An alert is bubbled up to human users only when the system can't handle it autonomously.

Data collection is a dominating pattern in edge computing, especially in projects that involve both cloud and edge. The most powerful aspect of this pattern is the ability to aggregate data from multiple contexts and extract additional value that can't be extracted from individual data sources. This ability also enables an important emerging concept – the *digital twin*. Digital twins are basically a digital simulation of a real environment and are used to control, simulate and optimize their physical counterparts. For this to work, constant and consistent data collection and uploading to the cloud is required. Azure Digital Twins is a good example of an implementation of the twin concept. It allows you to create a digital model of your actual entities such as buildings, rooms, sensors, etc. based on an open modeling language called Digital Twin Definition Language (DTDL). The twin instance in the cloud is connected with the actual assets and such allows users to bring any environment to life and apply business and data processing logic to it.

3.2 REMOTING PATTERNS

Back in the mainframe era, users attached to mainframe machines through *terminals*. All computation happens on the backend machine, and the terminal pipes user inputs and machine outputs between users and machines. This interaction pattern has been kept in traditional datacenters as well as on cloud. Administrators use terminals to attach to backend servers to operate the physical or virtualized machines.

As the bandwidth and reliability of public network improve over time, the interaction pattern is extending into more end-user scenarios, such as connected consumer devices, game streaming and other online collaboration scenarios.

Figure 3.2 illustrates a multi-user remoting pattern. In this pattern, real-world contexts are "teleported" into cloud to form a collective context. This allows physically scattered contexts to be joint together on cloud for remote collaboration scenarios such as online conference, collaborative editing and multiplayer gaming.

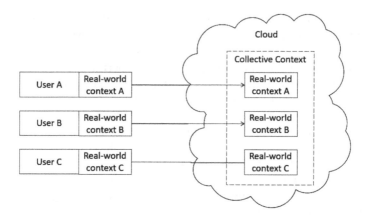

Figure 3.2 Remoting pattern.

3.2.1 Connected consumer devices

At the time of writing, the most representative connected consumer device is Google's *Chromebook*. Chromebook is a laptop or tablet that performs most of its tasks through a browser. Because most applications and user data reside on cloud instead of on the laptop itself, the laptop can be built with lower hardware configurations to provide similar functionalities with a more expensive laptop.

Chromebook is most successful in education, because it provides a cheaper alternative for students to access compute technologies.

In 2020, mobile device manufacturers such as Huawei are exploring to make fully connected devices such as fully connected laptops or cell phones. These devices offload all functionalities to cloud. They serve as pure input/output terminals (plus accessing peripherals such as cameras and microphone) to backend virtual instances. In other words, these devices are merely streaming terminals to actual compute instances on cloud, hence they can be made with much lower cost. The downside, of course, is that these devices need to be connected to function. This will be quite inconvenient when the network breaks. However, when reliable network connectivity becomes pervasive in the future, these devices are better choices for many reasons, such as cheaper price, better data protection, easier transition among devices, unlimited storage, longer battery life, lighter and slimmer design, and among other things, virtually immune to robbery or theft, which has been problems to owners of high-end mobile devices.

Game streaming is having a big impact on gaming industry. Game streaming services such as Google Stadia and Xbox Game Pass (formally xCloud) allow players to instantly access hundreds of game titles without having to download or install anything – everything is streamed from cloud. You can play game titles you own on PCs, tablets or cell phones from any place where a high-speed Internet connection is available. This kind of subscription-based gaming is rather different from the time when the authors grew up. At that time, acquiring a new game title was a big event, and players were more devoted to a single game title. Hence, games were much harder back then to give the developer enough challenges to perfect their gaming skills. And to get the maximum value out of a title, players tried to come up with new innovative ways to beat the game – no weapon upgrades, no deaths, high score, speed run and so on. Nowadays, gaming is more about an interactive experience than a skill challenge.

Cloud-based gaming also enables ambitious titles that are too big to be fit into any single console or PC. For example, Microsoft Flight Simulator 2020 constructs its virtual world using petabytes of data. This kind of massive game is unimaginable before the cloud era.

3.2.2 Online collaborations

In 2019, the world was hit by a highly contagious virus – COVID-19. At the time of writing, there's no effective way to treat the virus, but vaccines have been developed and are slowing down the spread, but yet it will still take some time to go back to normal. Like many others, authors of the book have been working from home for months, and the date to return to offices is out of sight. Restaurants, movie theaters and shops are mostly vacant. And when people go out, they wear masks and try to keep safe distances from each other. Fortunately, there aren't any zombies. Yet, it's a gloomy and lonely world.

The virus gives an unprecedented push in digital transformation. As schools closed and businesses shut down, online education and online meetings are the only way to carry on with business. As Microsoft's CEO Satya Nadella put it, "We've seen two years' worth of digital transformation in two months". Nadella was referring to the fact that the usage of Microsoft Teams, Microsoft's video conference solution, surged by over 400% in a couple of months. In the high school we volunteer in, everything switched to online formats – classes are held as online conferences, group discussions are moved to chat rooms, quizzes are provided as online surveys and homework is distributed as shared documents. And some traditionally in-person activities are transformed to online activities– such as using a remotely operated robot (as a surrogate of the actual worker – think of those sci-fi movies!) to organize store shelves, using drones to deliver packages and using a VR device to visit a temple.

A key characteristic of the remoting pattern is to push end-users' contexts to the cloud. This is what makes the pattern different from traditional web service patterns. With this pattern, a user's context is still logically attached to the user at the user's location. However, computation required by the context is pushed to cloud.

3.2.3 Cloud-based relays

Devices sometimes need to take ingress traffic. For instance, some surveillance cameras run a local media server that serves camera feed streams to attached clients. However, because devices often don't have public Internet Protocol (IP) addresses. And their private IP addresses are often from a network address translation (NAT) address space that isn't routable beyond the attached router. Further, for security concerns, inbound connections are often blocked on many local networks. All these make a service providing a publicly consumable service a difficult task.

This is where cloud comes in. Cloud endpoints have global presence. They can provide ingress endpoints to devices and egress endpoints to consumers. For example, in the case of surveillance camera, cloud can offer video ingress endpoints that take in video streams through RTMP protocol, and video egress endpoints that server up video in different formats, such as MPEG-DASH, Apple HLS, Microsoft Smooth Streaming and Adobe HDS.

Cloud-based relays allow devices with private IP address to connect with each other through outbound connections. When multiple devices form a temporary cluster for a task – such as a multiplayer gaming session or a chat room, one of the devices is elected to act as a server by certain criteria, such as having decent upload speed and low network latency. Some systems can leverage local area network (LAN) connections when possible. In such cases, devices talk to each other through peer-to-peer connections without going through the cloud. This provides faster connection speed and better privacy protection.

Cloud-based relay and cloud-based messaging can both be used to bridge devices from different locations. The main difference between the two is that the former focuses on joining two information paths, while the latter focuses on passing data packets between the two devices.

3.2.4 Multi-party computations

The ubiquitous connectivity provided by cloud enables many devices to collaborate with each other. These collaborations can be orchestrated by centralized servers or be self-organized by participating members. When we talk about multi-party computations, we are referring to specifically to the cases in which distributed compute resources work with each other to accomplish common tasks without centralized coordination.

One of the earlier successes in multi-party computations was peer-to-peer file sharing protocols such as BitTorrent. It was estimated that peer-to-peer network traffic contributed around 43%–70% of the entire Internet traffic. Instead of downloading large files from a few centralized hosts, BitTorrent used swarms of machines to jointly download and reshare files with each other. In the days when 56k dial-up modems were still cutting-edge, it was a wildly popular protocol for users to acquire and share large files such as videos, images and music.

Nowadays, one of the popular multi-party calculation scenarios is Blockchain. Blockchain allows participating parties, which don't necessarily trust each other, to reach consensus on transactions without central coordination. And transactions are recorded in immutable chains of blocks that can't be tempered after transactions are recorded, unless there is global consensus to prune the chain, discard all transactions from certain point and start over. The chain pruning may sound strange but it's an unfortunate necessity when the chain of blocks branches and only one of the branches can be kept.

The authors of the book believe Blockchain raises a valid scenario but provides an inefficient solution that is slow and over complicated. Distributed ledge and distributed consensus are two separate problems that can be solved relatively easily with existing replication and consensus algorithms. Bitcoin and other forms of cryptocurrency, which are a prominent application of Blockchain, is a very wasteful system that consumes ridiculously large amounts of global compute energy as the participating members repeatedly try to solve a digital puzzle as *proof of work*. In many ways, Bitcoins show the characteristics of a speculative bubble.

With the rising application of machine learning, cooperative training while preserving data privacy has gained lots of attention. Techniques like differential privacy and homomorphic encryption allow multiple parties to participate in collaborative training without disclosing private data. This is especially attractive to industries with sensitive data protection requirements and high demand on machine learning solutions such as medical industry and financial industry.

3.3 COMPUTE OFFLOADING PATTERNS

Due to various constraints such as cost, form factor, power supply and environmental aspects, edge devices are often limited in computing power, communication bandwidth and storage capacity. Such limitations require devices to offload complex compute tasks to local or remote servers that reside in fog or cloud.

The idea of offloading is simple – data is collected and processed by devices to generate decisions on actions to be taken (local loop). And when a device is overwhelmed, it offloads data processing tasks to an on-site server (remote loop) or a cloud/fog-based server (cloud loop) to make decisions on its behalf. However, when you take a closer look, you'll find more a few interesting and powerful variations, including bursting to cloud, multi-level offloading, dynamic offloading and peer delegation. Figure 3.3 illustrates these variations in a same diagram, highlighting

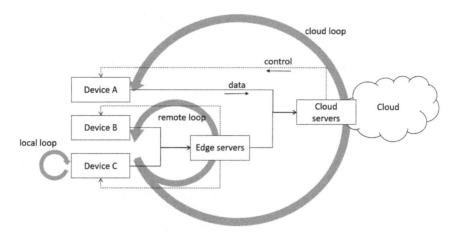

Figure 3.3 Compute offloading pattern.

feedback loops in different configurations. And we'll go through each of them in this section.

3.3.1 Multi-level offloading

The eastern port of Qingdao is one of the first fully automated ports. It's a so-called "ghost port" that doesn't need much lighting and there are barely any workers in sight. The sky-high cranes and the busy trucks are all operated in harmony by computers. The trucks are operated by AI drivers that carry out precise loading and shipping routes without colliding with each other. And the cranes unload crates from docked ships and place them on appropriate trucks with high precision without any coffee breaks (or tea breaks in this case). The port of Qingdao is a typical Autonomous IoT (AIoT) system, which is a combination of IoT and an autonomous control system (ACS). In such an AIoT system, some decisions are made locally to ensure a faster response time – such as avoiding an unexpected obstacle. Some other decisions can be offloaded to on-site servers – such as planning routes of the self-driving trucks. And infrequent and long-term decisions can be pushed further to cloud – such as schedules for predicative maintenance.

3.3.2 Bursting to cloud

In this pattern, compute happens on devices during normal workloads. When the workload has unexpected surges, the extra workload is shipped to cloud. Because workload spikes are only occasional, it doesn't make sense to keep a huge infrastructure on the cloud at all times, which incurs

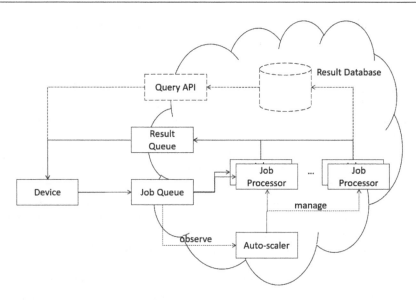

Figure 3.4 Bursting to cloud pattern.

additional cost. Instead, such systems often use an auto-scaler that spin ups compute instances only when necessary. Figure 3.4 illustrates a possible architecture of bursting to cloud. In this design, a cloud-based job queue is monitored by an auto-scaler. When a device is overwhelmed by increased workload, it sends some task to the job queue. Then, the auto-scaler decides on how many job processors to spin up based on the queue length. And when the queue is drained, it kills the processor to reduce cost.

There are multiple ways to send the results back to devices. Figure 3.4 shows two possible approaches – to send results through a result queue or to save results in a database and provide an API for devices to query the database.

The same cloud infrastructure can be shared to support multiple devices. However, there are few things you need to consider. First, you need to associate your jobs with device identifiers so that work results can be associated with the correct devices. Second, you may need to consider balancing job handling among devices so that a super busy device will not seize the entire job queue. And finally, you need to implement proper authentication so that the infrastructure is not abused by tampered devices.

3.3.3 Adaptive offloading

Adaptive offloading is an exciting pattern, in which devices dynamically decide on what jobs to be delegated to the cloud based on various conditions such as power level, environmental temperature, network condition, device load and many others. Because the pattern can be used to adaptive to endless combinations of environmental changes, it can lead to innovative solutions that leverage the best of both cloud and edge.

Early in 2020, most restaurants were closed due to the COVID-19 outbreak. Later, they were allowed to serve take-out orders only. Customers would call in to place their orders, and then drive by to pick up their food. This was when my colleagues and I started to prototype an automatic parking lot food delivery system using Boston Dynamics's robot dog – Spot (because why not, right?). The system used a camera to capture cars driving to the parking lot. Then, it used a computer vision system to read license plate numbers. It matched the license numbers with orders and sent related information to the Spot, including the order to be delivered, characteristics of the car (color, type, rough location, etc.) and how long the customer had been waiting. Finally, a human worker loaded the food to a small basket mounted on Spot, and Spot delivered the food to the hungry customer without direct human contact.

By default, images were sent to the cloud for analysis. This is because open space license plate detection was quite difficult. We used a multi-stage pipeline that found cars using object detection, and then located license plates using computer vision, and then recognized texts using optical character recognition (OCR). And we had to do fuzzy matches between OCR results and known license plate numbers. This entire pipeline ran on cloud at about 3 frames per second on CPU.

The system was also integrated with a network monitoring system. When the network condition degraded, the system switched to images with lower resolutions to keep up the frame rate. And if the network was completely broken, it switched to a local pipeline with lightweight object detection framework and less sophisticated OCR detection to provide continuous service on the local device – which is a Windows tablet – when the network was out.

In our case, the application proactively switched between different pipelines (high-resolution, low-resolution, local) based on network telemetries. When we discuss capability-oriented architecture (COA) in Chapter 7, we'll examine how this adaptiveness can be pushed down to the infrastructure layer without any adaptive code in applications themselves.

3.3.4 Dynamic delegation

When one of the doors of my smart fridge is kept open for too long, a message will pop up on my TV telling me the fridge door is open. This is a simple use case of a powerful pattern – dynamic delegation.

Dynamic delegation allows devices to discover capability providers on the same network or over the cloud, and delegate tasks to more appropriate or more powerful devices. In the previous example, displaying a message on the TV is definitely a more effective way to display a notification to end users – ordinary people spend much more time watching TV than staring at a refrigerator. Another scenario of delegation in my house (as mentioned above, I'm staying at home a lot nowadays) is to play my Xbox console games on my Surface Duo. In this case, the console is doing all the complex calculations and rendering, and the game video is streamed to my phone through the network.

In both cases, a device uses another device's capability to enhance its reach to provide fresh user experience. The device functions fine without any peer devices. However, when peer devices pop up online, the device can discover them (within the same ecosystem – Samsung for the fridge case and Microsoft for the Xbox case) and leverage their capability to deliver additional functionalities.

This discovery process can be done in several different ways. First, devices owned by the same user can be linked when the devices are registered under the same user account. This linkage is not bound to any networks and the devices can remain linked over the globe. Second, a device may be configured as a *hub* device that publishes a rally point for devices to connect with each other over a local network. And finally, devices can broadcast capabilities (by sending messages over User Datagram Protocol (UDP), for example) on a local network.

If you think about it, delegation happens between devices and cloud all the time. Essentially, many cloud service calls form devices can be considered devices delegating tasks to cloud – to recognize a face, to transform a file, to detect an anomaly and so on. The difference between these delegation calls and regular web service calls is that these calls are meant to assist the device to accomplish a task within the local context. For example, using a cloud service to recognize face is to help a local camera to identify a person. On the other hand, watching online video is not a delegation because the activity is not originated from a local context.

COA natively supports dynamic delegations. We'll discuss the COA's view on dynamic delegation in Chapter 7.

3.4 SUMMARY

This chapter reviews three groups of patterns that connect edge devices to cloud, including data collection patterns, remoting patterns and compute offloading patterns. Data collection patterns focus on gathering and extracting value from distributed device data. Remoting patterns allow user contexts to be transmitted and handled in remote locations. And finally, compute offload patterns allow devices to delegate compute tasks to cloud-based compute resources or other peer devices. The interconnection between devices and cloud is so strong in these patterns that it leads some to believe edge computing is all about connecting devices to the cloud.

At the time of writing, moving data and compute to cloud is still the mainstream edge computing pattern in the industry. However, we've started to see data and computing moving in the other direction. Although we believe edge-to-cloud patterns will still go strong in the years to come, we also believe cloud-to-edge patterns present some new and exciting opportunities. Especially, we see the cloud-to-edge patterns as enablers of an era of ubiquitous computing. In the next chapter, we'll examine some of the emerging patterns that disperse computing toward edge so that compute can happen right in the context of edge. And we believe that's the true meaning of edge computing.

Chapter 4

Cloud to edge

A few years back, I had eye surgery in California.

Before the operation started, I noticed that all knives and needles were mounted on a rack. I asked the doctor why that was the case. She told me that the rack provided a steadier hold than human hands, so it was safer for my eyes.

"Also, the rack is connected to an earthquake warning system", she continued.

"Why?" I asked.

"California has many small earthquakes. The warning system can tell me when an earthquake is about to happen", she explained.

"How far in advance?"

"About 5 seconds".

"5 seconds? That's not a lot of time!"

"Well, it's enough time for me to retract all knifes from your eye", she smiled.

This is a different edge application pattern than what we saw in the previous chapter. In this case, the device (the rack) responds to a notification from cloud and performs the necessary operations to retract all knives and needles safely and swiftly. In other words, this pattern pushes data to devices to trigger actions in a user's context. Of course, on the other end of the system, some devices can collect ground vibration data to cloud, following one of the data collection patterns discussed in the previous chapter.

In this chapter, we'll review some of the cloud-edge patterns that push data and operations to connected devices, starting with patterns that aim to accelerate cloud services.

4.1 EDGE ACCELERATION PATTERNS

In these patterns, edge is indeed "edge of the cloud". It's used to provide a better user experience by caching cloud data and performing some user-specific calculations. Further, because a large portion of user workload can be handled directly on edge devices, these patterns help the cloud-based server to scale to a much larger scale (up to about ten times) using the same amount of cloud resources.

4.1.1 Smart CDNs

Content delivery network (CDN) is the classic cloud acceleration service that uses a geographically distributed network of points of presence (PoPs) to deliver content to end users. CDN is an effective website accelerator because static content such as images, style sheets and Hypertext Markup Language (HTML) templates can be served directly from PoPs instead of requiring complete roundtrips to the original server.

Traditionally, CDN serves as a distributed cache of origin sites. In recent years, CDN vendors start to offer service hosting directly on their infrastructure. For example, instead of publishing a website and then enable CDN in front of it, you can deploy your website directly to CDN nodes without setting up an origin server. For lack of a better name, we'll call these CDNs "Smart CDNs" for now.

Vercel (previously ZEIT, see https://zeit.co/smart-cdn) offers rapid website deployments using a Smart CDN infrastructure. It allows you to publish static websites, functions and server-side pre-rendering on their edge location hosts. It offers 23 global CDN regions with 99.99% availability guarantee. It's used by some well-known sites such as Airbnb, Auth0, Twilio and TripAdvisor. They built (together with Google and Facebook) an open-source Reactive framework named Next.js. One interesting feature of Next.js is to progressively generate static contents that can be cached on edge nodes. This allows both static content and dynamic contents to be effectively cached on the Vercel global edge network. Next.js is one of the static site generators out there, such as Eleventy, Hugo, Docusaurus, Nuxt, Gatsby and others. These generators generate web application as pre-built static pages. This architecture of using reusable application programming interfaces (APIs), client-side JavaScript and markup content is also called a Jamstack architecture. And the sites built using the architecture are called Jamstack sites or Jamstack apps.

Netlify Edge is one of the original creators of Jamstack. It also offers a platform (see https://www.netlify.com/products/) that allows you to publish sites directly to their edge presence without an origin server.

4.1.2 EdgeWorkers

Akami uses EdgeWorkers (https://developer.akamai.com/akamai-edge-workers-overview) to deliver better web experiences by allowing running JavaScript functions at edge. Akami EdgeWorkers is not designed for generic compute on the edge (comparing to the edge function pattern below). Instead, it's specifically targeted at improving experiences of consuming the backend service. It defines a simple API through which you can plug in JavaScript snippets into your ingress or egress traffic on the edge for scenarios like A/B testing, filtering traffic and customizing responses. However, it seems the long-term goal of Akami EdgeWorkers is to evolve into a full edge function platform.

4.2 CLOUDLET PATTERN

In this pattern, cloud platform vendors integrate with the 5G network infrastructure to bring their cloud services closer to the consumer. In this pattern, a cloud vendor deploys a small-scale datacenter, which often uses the same cloud datacenter design and hardware, directly into 5G partner's network infrastructure. This arrangement is sometimes referred to as *cloudlet*.

Cloudlet is one materialization of multi-access edge computing (MEC), which is a concept to bring compute, storage and analytics capabilities closer to the end user. It's considered one of the foundational technologies of 5G. In this book, we examine edge computing patterns from an application's perspective, while MEC is often discussed in the context of infrastructure and networking. Hence, we'll continue using cloudlet as the pattern name in this book.

Controls and user plane separation (CUPS) plays an important role in realizing MEC. It separates user data flow from the control plan data flow and allows user data flow to take a much shorter path to finish the roundtrip. We'll present a few examples in the next few sections. Because 5G is mobile network infrastructure we need to account for mobility and network capacity changes for the end user. The network API standard described as part of ETSI TS 129 522 V15.2.0 is aiming to expose the network conditions to the application, and such provide and improve the end user experience.

4.2.1 AWS Wavelength

Amazon Web Services (AWS) Wavelength makes popular AWS services available at Wavelength Zones so that applications can leverage popular AWS services with single-digit millisecond latency over 5G networks. These services include Amazon Compute Cloud (EC2) instances, Amazon Elastic Block Storage (EBS) volumes, AWS Elastic Container Service (ECS), Amazon Elastic Kubernetes Services (EKS), AWS Identity and Access Management (IAM), AWS CloudFormation and AWS Autoscaling.

Wavelength reduces latency by minimizing number of hops the devices must go through to connect to these services. It's designed for scenarios that require high bandwidth and low latency, such as multiplayer game streaming, MR/AR and automatic cars. It partners with telecommunication providers such as Verizon, Vodafone, KDDI and SK Telecom to allow mobile users to connect to AWS services directly through mobile networks. Figure 4.1 compares the latency between accessing cloud and accessing cloudlet. In the case of accessing cloud, a mobile device needs to be connected through cell towers to content service provider (CSP) infrastructure, and then to transit/peering point into the Internet, and eventually to the cloud. This chain of hops generates about 100 ms roundtrip latency. In comparison, because cloudlets run directly on CSP's infrastructure, a significant portion of the path is cut, reducing the roundtrip latency to 10 ms range.

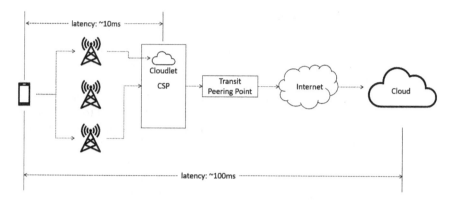

Figure 4.1 Cloudlet vs. cloud.

Furthermore, because routing to the cloudlet is much better controlled and predicable than Internet routing, users experience less jitter, which helps to build a consistent and predicable user experience.

4.2.2 Azure Edge Zones

At the time of writing, Azure has just announced Azure Edge Zones preview, which delivers functionalities that are similar to what AWS Wavelength offers based on Azure Stack Edge platform. Azure Stack Edge is a hardware-as-a-service (HaaS) solution that brings selected Azure services to a small server rack form factor. I remember running a prototype rack in my own office a few years back. It was quite loud and too hot for an office setting, and I got complaints from my neighbors – especially the guy whose office is across the hallway because I had a powerful fan that blew hot air out of my office right into his. I had to give up the rack after a few weeks. And I believe it's now happily running in some properly cooled (and sound-insulated!) labs.

Virtualization is a pivotal technology that enables MEC scenarios. Specifically, virtual network functions (VNFs) separate network tasks (such as DNS, NAT, caching and firewall) from proprietary, dedicated hardware and put them into commodity servers or virtualized servers. This gives network carriers great flexibility in designing their software-defined networks (SDN) topologies that allow CUPS and private networks between CSP and clouds over the same physical network infrastructure.

4.3 EDGE FUNCTIONS PATTERN

The cloudlet pattern pushes cloud infrastructure toward edge so that developers can use the same set of cloud capabilities on telecom infrastructure or on edge. This kind of cloud-edge infrastructural parity certainly helps to create certain consistency. However, developers often prefer to work on a higher abstraction level (such as platform as a service, or PaaS) instead of working directly on top of infrastructure. Furthermore, replicating PaaS behaviors doesn't necessarily require replicating the underlying cloud infrastructure, which is quite complex.

Functions is a popular programming pattern for reactive, distributed microservice applications. In theory, you don't need a complex infrastructure to support functions – as long as you can host a computing unit that can take some input and generate some output, you've got a basic function runtime. The edge function pattern focuses on bringing function runtimes to edge. These functions are often part of a larger system. They bring selected tasks closer to end users to provide faster performance, while leaving more complex tasks back on cloud.

Figure 4.2 shows a typical architecture of the pattern: A function host is deployed on CSP infrastructure or on-premises servers to host function units. The host supports specific programming patterns and frameworks that are chosen by the vendor. For example, some hosts support Docker containers, while others support a specific function runtime such as AWS Lambda and Azure Functions. Because a function host is lightweight, it can also be deployed directly on capable devices to provide even faster response times. The function units get and receive messages through a messaging backbone. And they expose functionalities to devices through messaging or standard protocols such as Hypertext Transfer Protocol (HTTP) and gRPC. The messaging backbone is often hosted on cloud. To enable offline communication, some function host can act as a message bus itself. When a group of devices is disconnected from the cloud, they can elect one of the hosts to act as the local messaging backbone to support continuous communication among the devices. The control plane that manages the function hosts is often kept on cloud to manage distributed hosts on edge infrastructure.

Functions can be *blocking* or *non-blocking*. Blocking functions usually directly return the execution results to the caller. Non-blocking functions work in a fire-and-forget pattern. Once the function is invoked, you may

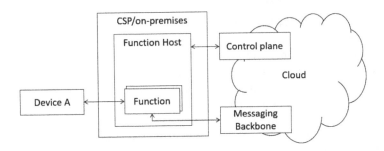

Figure 4.2 Edge functions.

or may not get a result back from a separate channel (such as a message queue). Most functions are stateless, which means they don't keep any local states. There are stateful functions but state management, especially high availability state management, requires additional infrastructural support, which is often missing on the edge.

Next, we'll examine a few edge function runtimes.

4.3.1 AWS IoT Greengrass

AWS IoT Greengrass (https://aws.amazon.com/greengrass/) allows you to run AWS Lambda functions (as well as Docker containers) on edge devices, with or without connection back to the cloud. On the other hand, when the devices are connected, you can send data back to AWS cloud, or use various *connectors* to connect to AWS services.

AWS Lambda is a programming model support by AWS cloud. AWS offers serverless hosting of Lambda functions on cloud. And you can use the same programming model for edge workloads. Having a consistent programming model across cloud and edge is beneficial in several ways. First, it allows code reuse across cloud and edge. Second, it reduces tech stack complexity to reduce chances for errors. Third, because you have the same runtime on both cloud and edge, you can develop and test your Lambda code on cloud without having to connect to an actual device.

Devices are configured into *Greengrass groups*. A group is defined and configured by a *Greengrass core* residing on cloud. The Greengrass core defines the topology of physical devices. And it's materialized onto physical devices through an agent.

Devices within the same group can communicate with each other. This is done by an on-device gateway that passes messages among devices (via Message Queuing Telemetry Transport (MQTT)) without going through a cloud-based message backbone. Greengrass also ships with necessary security measures to secure cross-device calls, to enable AWS service consumptions and to manage device identity.

4.3.2 Azure IoT Edge

Azure IoT Edge runs IoT Edge modules on edge devices. The Azure IoT Edge runtime (runs as a Docker container) connects to Azure IoT Hub (as a messaging backbone) and manages modules running on the edge device. It also has an IoT Edge Hub (note the subtle differences in word combinations) that acts as a local message bus among modules for offline scenarios.

An IoT Edge module is packaged as a container image. It can be Azure Functions, Azure Stream Analytics *jobs*, Azure Machine Learning *modules*, or (1st-party or 3rd-party) custom code using the module software development kit (SDK). Azure IoT Edge also allows management by a Kubernetes control plane through virtual Kubelet. The idea is to allow managing hybrid apps that span cloud and edge or multiple Kubernetes clusters. It is also worth pointing out that Azure Functions are portable and as such can also run as standalone and independently from IoT Edge on edge devices.

4.3.3 Apache OpenWhisk

Apache OpenWhisk is an open-source function runtime for serverless platforms. It allows you to write functions (named *Actions*) in a bunch or programming languages (Node.Js, Swift, Java, Python, etc.) to respond to events (through *triggers*) or HTTP requests. You can also declaratively attach multiple *rules* to the same trigger to trigger multiple actions upon a single event. As the runtime is also containerized, you can in theory deploy and host the runtime anywhere containers are supported.

IBM offers an IBM Cloud Functions built on top of OpenWhisk. Cloud Functions is designed as part of its serverless platform (which is part of IBM Bluemix cloud platform) but supports edge function workloads as well. IBM also has an Edge Application Manager, which is designed to manage applications on edge at scale.

As shown in Figure 4.2, an edge function system needs to deploy and manage function codes in many scattered function hosts. Managing a large, distributed device fleet is complex. We'll use the next section to discuss elements in managing a scalable edge function platform. Although the discussion focuses on edge functions, many of the elements are applicable to other types of managed runtimes such as edge-based PaaS, which we'll discuss in Section 4.4.

4.3.4 Function host management at scale

A scalable edge function runtime involves some key components:

- Function hosts. These are the physical or virtual devices that provide the Operating System to support a function runtime. In a *zero-trust* environment, a function host must prove its legitimacy (through attestation, for instance) before it can be trusted with function workloads. It also needs to have a valid registration with the cloud-based control plane to establish its identity.
- Function runtime. Function runtime is often a combination of system services (or agents) and supporting libraries. The services are responsible for maintaining communication pipelines (such as triggers) between function code and the edge environment. They also supervise local function processes and take necessary recovery actions to ensure function availability. Another common task of these services is to keep the runtime itself up to date by applying updates and patches. Finally, these services also provide basic utilities such as logging and sending heartbeat signals.

- Function packages. These packages carry the actual function code. They can be platform-specific packages, Docker images, and sometimes virtual machine (VM) images for hypervisor-based edge function runtimes. These packages use SDKs provided by the corresponding function runtime to hook up with the message pipelines provided by the runtime. Most function runtimes support running multiple function packages at the same time, assuming these packages belong to the same tenant (i.e. not hostile to each other). Configurations of these functions are maintained on the cloud-based control plan. These configurations are brought down and applied by the function runtime.
- Data pipelines. Data pipelines allow functions to interact with the device environment. These pipelines are often message-based through a local or a remote (or a combination of both) message bus. Function runtimes usually support extensible *connectors* that facilitate connections to and from popular systems such as Kafka.

To ensure everything is laid out and configured properly on all devices, an edge function system often supports the concept of a *device twin*, which defines the software stack as well as the desired configuration state. Then, devices are grouped together by either static grouping or dynamic filters. And finally, the control plane coordinates with edge agents to drive the device state toward the desired state. This configuration roll-out process is eventually consistent. This means during a large-scale roll-out, there will be devices running the old code (and potentially using an older message format) while others are running the new code. This requires the backend to be able to handle multiple message versions and/or APIs at the same time.

Finally, when you work with a large number of devices – such as hundreds of thousands of devices – you need to consider not overloading a single data pipeline. In such cases, you may want to load-balance your devices across multiple data pipelines.

Edge functions satisfy the needs of many simple scenarios. And because edge function runtimes are often quite lightweight, it's suitable to be deployed to large device fleets at relatively low cost. However, for more complex scenarios in which you need to run multiple collaborative components on devices, you need a more sophisticated platform, such as cluster compute resources. We'll spend the next chapter explaining how to use Kubernetes to manage a device cluster on edge.

You may wonder why not run a full-scaled PaaS platform on edge instead of running just simple functions. We'll use the next section to discuss why running PaaS on edge is a questionable idea.

4.3.5 Edge PaaS paradox

PaaS on cloud fulfills two key cloud promises – elasticity and availability. PaaS separates applications from the underlying infrastructure and allows compute resources to be dynamically allocated to support the applications. At this point we need to decide between application and infrastructure level elasticity.

You can enable application elasticity by scaling out, meaning adding more instances of your services or containers on existing compute resources, but at some point, your application may need additional compute. PaaS enables elasticity at an infrastructure level because you can acquire additional compute resources at any time to scale out your application and release unused resources when you don't need them anymore. PaaS also enables availability by running redundant instances of your applications so that when some of the application instances fail, there are still other application instances to handle user requests.

Almost intuitively, people believe bringing PaaS to edge is a natural choice, as it brings all PaaS benefits to the edge and ensures consistency across cloud and edge. However, is it that simple?

Let's examine the two value promises of PaaS – elasticity and availability. The foundation of elasticity is the availability of a (large) resource pool; and the foundation of high availability is redundancy. In other words, to take full advantage of PaaS, you need stand-by resources, which is a luxury that is too expensive to be afforded on edge. On the other hand, even if you had enough extra resources, you don't have the same flexibility in workload scheduling as you have on cloud. This is because edge computing is context-aware. You can only relocate workloads within the physical contexts. For example, you have 20 devices running face detection on ten cameras, with two devices assigned to each camera for redundancy. When both instances for a camera fail, you can't failover to any of the other 18 devices because they were attached to different camera feeds.

PaaS does bring consistency in terms of applications model; programming model and toolchain supports when it's deployed on both cloud and edge which highlights another aspect of the complexity of providing PaaS on the edge. In addition to elasticity and availability PaaS also promises no downtime application runtime upgrades. For example, when a security update for the. NET runtime on Azure App Services is applied the user application remains available. This can only be accomplished through compute resource redundancy and stateful compute.

The ultimate goal is to develop an application that runs on the cloud and on the edge. While this might be a valid use case for server class edges it is rarely realistic, at least not without application modifications, for applications targeting low capability edges with the exception of some AI models. PaaS on the edge is nice idea, and it does have certain usages in limited scenarios (in which edge compute resources are abundant). However, you need to be aware of the limitations as well as associated cost running PaaS on edge.

4.4 CLOUD COMPUTE STACK ON EDGE

Cloud datacenters are organized differently from how traditional datacenters are managed. Traditional datacenters often pin workload on specific servers – such as database servers, file servers, mail servers, etc. On the contrary, cloud datacenters separate workloads and hosts. Cloud compute resources are organized into a shared resource pool, on which different workloads are scheduled. Over the years, cloud platforms have pushed automation and energy efficiency to the extreme to offer sustainable, reliable and economical cloud services. These characteristics are quite attractive to traditional datacenter operators, especially to telecom companies who seek to offer cloud capabilities on their infrastructures.

On July 9, 2020, two huge white metal cylinders surfaced by the shore of Scotland's Orkney Islands, after laying on the seafloor 117 feet below surface for over 2 years. Inside the cylinders are racks of Azure servers that have been operating without any human intervention in the watertight submarines filled with nitrogen. This is project Natick, an experiment to test if autonomous datacenters can be reliably operated offshore in the ocean. The project worked out great. First, because of less erosion in oxygen and human error, server error rates were reduced by 50%. Second, because the system was essentially cooled by ocean current, the energy consumption was reduced to minimum (with no apparent effects on marine lives around the tanks – I've seen the live footage and the fish looked happy, at least). Third, most active economic circles of the world are near the seashore, deploying datacenters directly offshore provides a direct and fast connection to these underwater centers.

Project Natick proves that deploying small-scale, autonomous cloud datacenters is feasible. However, running datacenters underwater still feels futuristic. In the next few sections, we'll examine some other solutions that are not as extreme.

4.4.1 AWS Outputs

An AWS Output is a server rack that is managed by AWS but physically deployed at on-premise facilities or co-location spaces. When servers fail, AWS is responsible to fix or replace them, so that the customer who deploys the Output isn't burdened by infrastructural management – a major benefit they get using the cloud. Output supports a subset of AWS services, including EBS, EKS, Amazon Relational Database Service (RDS) and AWS Elastic Map Reduce (EMR).

The goal of running cloud infrastructure on edge is to bring consistency in both development cycles and service operations, because developers and operators will be able to use the same set of libraries, programming patterns and toolchains to design, develop and manage their services that need to be kept on-premises for various reasons such as privacy concerns (such as medical imaging and patient record management), latency requirements (such as real-time manufacturing control and ERP transaction processing) and volume of data (such as 3D modeling and video processing).

The hardware you get in an Outputs rack is identical from what AWS runs in its own datacenters. The machines are built on the same AWS infrastructure architecture, and they have their own individual power supplies for better AC/DC conversion efficiency just like they are in AWS datacenters. AWS manages and monitors Outputs sides just like managing any other AWS region, which we will return to in a moment. And you use the same set of tools and APIs to interact with the infrastructure. So, did AWS really cut off a piece of AWS and hand it to you? Unfortunately, the answer is not that straightforward.

As we can easily imagine, operating an AWS region is not an easy task, which requires supports of hundreds or even thousands of highly available microservices. Naturally, you may wonder how much of the rack's resources are used to operate the control plane to keep your "local cloud" running. And as everything is running on the same rack, how do you ensure control plane availability with limited resources?

These are all great questions. And it turns out, AWS uses a different approach – it keeps the control plane on cloud. An Outposts deployment is connected to the nearest AWS region through either AWS Direct or VPN. This creates a reliable private connection between the Output deployment and the AWS region. For instance, you can join Outposts EC2 instances to the same virtual network (VPC) where your on-cloud EC2 instances reside. Through the same private connection, AWS connects the control plane to your Output servers, as shown in Figure 4.3. The figure shows that the cloud control plane, which is deployed across multiple availability zones for high availability, is connected to an Output edge device through a group of Output proxies.

The edge device is comprised of Nitro cards, which are designed for I/O acceleration. Nitro cards is one of the three key aspects of AWS infrastructure architecture: Nitro cards that provide I/O acceleration, Nitro security chips that provide hardware root of trust and Nitro hypervisor that provides infrastructure virtualization. The Nitro security chip plays an

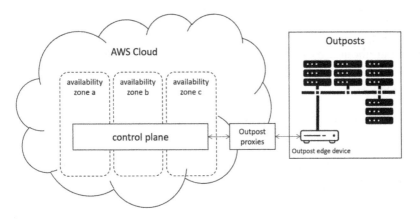

Figure 4.3 AWS Outposts.

important role here because the control plane in this case is operating on machines that are outside of AWS centers, which can reinforce layers of physical security to protect the servers. Now as the servers are installed on-premises, AWS must make sure none of the hardware, firmware or host have been tempered in order to sustain the strong security guarantee.

4.4.2 Azure Stack

Azure Stack was originally introduced to bring Azure infrastructure and services to on-premises datacenters. Later, the product was evolved into a family of offerings, including Azure Stack Hub, Azure Stack HCI and Azure Stack Edge. Azure Stack Hub is the original Azure Stack. It offers Azure infrastructure and PaaS services on on-premises datacenters. Azure Stack Edge is for offline compute and data transfer scenarios.

Azure Stack HCI (hyper-converged infrastructure) is bundled into Windows Server 2019 and runs on hardware from validated partners. It offers the fundamental HCI platform including Hyper-V, software-defined storage (SDS) and SDN for disconnected, on-premise workloads. Azure Stack HCI requires minimum of two servers, which can be connected through a switch or cross-over cables.

I had the opportunity to play with an early version (pre-alpha) of Azure Stack a few years back (way before Output was announced). At that time, the idea is to essentially package the entire Azure software stack as an installable package. You need to set up your own servers and networks, and then install Azure on your servers. Then, you get a totally disconnected private Azure with everything – resource manager, portal, local services such as VMs and storage. I don't remember the required machine specs but those were some powerful servers with lots of memory and disk spaces. With the help from the team, I was able to install and configure it within 2 days.

Later, I shipped the server to San Francisco for a demo in a conference. Two nights before the demo, they asked us to clean up backstage before the big show, so I had to roll the server rack (a small one, in the size of a mini fridge) on a food cart down the San Francisco streets back to my hotel room at about 2 o'clock in the morning. I covered the $30K server with a tablecloth. And with the way I dressed, I blended in with wandering homeless people, who were also pushing their carts. It was certainly a unique experience.

Fast forward a few years, nowadays to get an Azure Stack Hub, you need to work with one of the certified original equipment manufacturers (OEMs) to get one. It is not the "white glove" experience you have with AWS Outposts, but you can certainly save some configuration time. A key difference between Azure Stack Hub and AWS Outposts is that you can run Azure Stack Hub in a completely offline mode – because you have your own copy of the control plane. Yes, you spend the extra resources to run the control plane yourself, but you get the true offline scenario in return – such as when you are running your datacenter on a cruise ship (before the SpaceX's satellite Internet becomes a thing).

Azure Stack Edge is a fully managed device by Microsoft. You can buy the 1U server and plug it to your server rack. Azure Stack Edge also brings some Azure services to the edge. However, it uses a rather different architecture. Instead of trying to replicate cloud infrastructure, it relies on containers (which we'll talk about more in the next chapter) to ship containerized Azure services to the edge, including Azure Cognitive Services, Azure Machine Learning, Azure Stream Analytics, Azure Functions, SQL Database Edge and Event Grid.

In 2020, Microsoft also announced an Azure Space offering, which combines satellite-based connectivity and Microsoft Azure Modular Datacenter (MDC) to take cloud compute capabilities to remote edge – even the Moon or the Mars.

4.4.3 OpenStack

The goal of OpenStack is to create an open-sourced cloud-computing platform, focusing on infrastructure as a service (IaaS). Founded by Rackspace Hosting and NASA, the platform follows an open philosophy that advocates for openness in every aspect – design, development, community and of course source code.

OpenStack is not one thing you buy or one service you subscribe. Instead, OpenStack is comprised a number of projects (or components) that deliver necessary capabilities to support a public or private cloud platform – network, storage, compute, application orchestration. Figure 4.4 illustrates some of the OpenStack key components, their functionalities and relationships with each other.

Table 4.1 provides a side-by-side comparison between OpenStack, AWS and Azure. If you are familiar with either AWS or Azure, you can roughly map their services with corresponding OpenStack projects in the table.

OpenStack contains a growing list of projects. It will be quite an expensive process to try to install, configure and operate these projects yourself. There are few vendors that offer managed OpenStack services, including IBM Bluemix, Cisco Metacloud, Ubuntu BootStack and a few others.

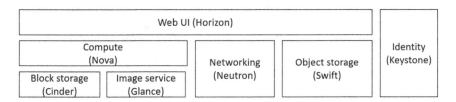

Figure 4.4 Key OpenStack projects.

Table 4.1 Comparison between OpenStack, AWS and Azure

OpenStack	AWS	Azure
Barbican	AWS HSM	KeyVault
Ceilometer	Cloudwatch	Azure Monitor
Horizon	AWS Management Console	Azure Management Portal
Heat	Cloudformation	ARM
Keystone	IAM	AAD
Mistral	SWF	Logic Apps
Neutron	VPC	Virtual Network
Nova	EC2	VM
Swift	S3	Storage
Trove	RDS	SQL Database, Cosmos DB, PostgreSL, MySQL, MariaDB, Redis
Zaqar	SQS	Service Bus, Event Grid, Event Hub

These are cloud providers that offer managed OpenStack deployments. They don't aim at providing edge-based deployments.

OpenStack took in StarlingX (https://www.starlingx.io/) in 2018, as it took in some other infrastructure projects that are in parallel with OpenStack architecture. If my memory serves me right, StarlingX initially intended to provide a hardened OpenStack platform on edge. Later (version 2.0, I think) it pivoted heavily toward Kubernetes and started to focus on managing container-based workloads. And OpenStack in this context becomes a host manager that manages physical and virtualized hosts for bare-metal containers such as Kata containers.

At the time of writing, pushing cloud infrastructure to the edge has limited success. There are many factors at play. But I think one important factor is the misconception of the value of cloud-edge parity. Cloud uses a generic compute pool to serve for many tenants and applications; while edge often serves for a specific tenant or even a specific application, which is often tied to specialized hardware. In such cases, the way that cloud operates doesn't work well, and the elasticity promise of cloud doesn't apply.

On the other hand, Kubernetes is becoming a popular infrastructure platform for containerized applications on edge because it provides a consistent hosting environment on both cloud and edge. This proves that the parity at application level seems to make much more sense than parity at infrastructural level. We'll discuss Kubernetes on edge in the next chapter.

Google Anthos is an excellent sample of application-level parity. It gives you a consistent hardened Kubernetes environment on both cloud and edge. Given Kubernetes originated from Google, it can be imagined the project has some immediate credibility upon release. At the time of writing, it's still early to judge if it's a business success so I'll leave that to your own discovery.

4.5 EDGE NATIVE COMPUTE STACK

Edge is not the edge of cloud. Edge represents an open, dynamic and het-erogeneous compute plane. Because of the broad range, edge covers a great spectrum of hardware and software. And you can find pretty much all software design patterns applied at different scales in edge computing. In this section, we'll break away from cloud and focus on those application patterns that are born on edge and grow on edge. We picked a few sys-tems that represent several different trends of thought – to uplift existing infrastructure, to design an edge native framework and to make untrusted parties work together.

4.5.1 Uplifting on-premises datacenters

Back in the mainframe age, compute resources were quite centralized. Then, as the personal computers arose, compute became distributed but pretty much isolated. Later, the Internet brought machines together again and popularized client/server and peer-to-peer architectures. Then, e-Business became a thing and compute was pushed toward the server with browser/ server architecture. Then came Salesforce, who popularized the idea of software as a service that separated service consumers and service opera-tors and switched people's mindset from license-based mode to subscrip-tion-based mode. The cloud revolution started by AWS further separated infrastructure operators from service operators with IaaS. And then, tech-nologies such as OpenShift, Docker, Swam and Kubernetes started to seek footholds in the PaaS era. Finally, 5G got lots of attention as its ubiquitous, high-bandwidth connections blurred the traditional boundaries between server and client, cloud and edge and opened a whole new set of possibili-ties and scenarios. If you have been living under a rock, now you are pretty caught up with what's been happening in the last 30–40 years in the com-mercial software industry, at a very high level.

As you can see, the software industry has been evolving fast and dras-tically. Compute resources are condensed, dispersed and then condensed again. And now we are in another wave of distribution of compute resources. And in this new wave, something interesting happened: some compute resources that are left out during the cloud transition are seeing a new wave of compute tasks coming back to them before their life span expires. These resources were deemed obsolete because their workloads had shifted to cloud. However, now they need to take on the new role of supporting edge computing for the years to come.

However, this time around, people are seeking to organize these compute resources in a more modern way. For example, many telco companies want to bring agility and cost-effectiveness to large number of *central offices* (CO) they own (AT&T alone has over 4,700+ central offices). There's actu-ally a term for this – CORD, which means Central Office Re-architecture

Figure 4.5 CORD high-level architecture.

as a Datacenter. The goal of CORD is to bring a cloud-inspired datacenter architecture to these central offices so that each of the offices can operate as a mini edge cloud. The focus of CORD has been enabling telco operators to refactor their central offices and push the central offices functionalities toward the consumer. However, the work also lays the foundation for hosting third party applications, especially containerized applications on Kubernetes.

Modernizing these central offices is challenging, because the infrastructure is the result of tens of years of accumulation of heterogeneous, proprietary hardware that is expensive to maintain or reprogram. The strategy of CORD is essentially break down the field office infrastructure into small, software configured pieces – such as SDN, VNF and VM – and reassemble them together to support modern workloads, or to provide edge access points to cloud platform, which is referred as access-as-a-service. Figure 4.5 shows a (very) high-level architecture of CORD. OpenStack and Docker provide VM and container management; and Open Network Operating System (ONOS) provides networking functionalities such as virtual networks and virtual routers. And finally, services are built on top of this infrastructure.

Efforts like CORD present an opportunity to construct a "edge native" cloud that is closer to end users and has direct integration with (5G) mobile networks. If executed correctly, such clouds could impose a real threat to classic clouds, especially in consumer-focused scenarios such as media streaming, smart homes and small businesses.

CORD represents an infrastructure-focused approach to build an edge native compute stack. It decomposes traditional datacenter resources into reusable components through virtualization. This layer of virtualization decouples workloads from the underlying infrastructure, hence allowing workloads to be orchestrated on top of a shared compute resource pool, just like how cloud does it. So, we can call this design cloud-inspired edge compute stack.

4.5.2 Template-based compute stack

Deploying an edge computing solution is complex. Many companies, products and frameworks aim to streamline the deployment process to get their customers started on the edge computing journey. How would you streamline a complex process? Providing a wizard or a template seems to be a straightforward solution. And this is the approach taken by several systems. For example, Azure IoT Central allows you to create applications based on four templates: retail, energy, government and healthcare. The templates allow you to quickly bootstrap your applications with simulated devices, which are configured with end-to-end pipeline to collect, analyze and present sensor data. Then, you can customize your application toward your specific needs.

Designing and implementing a comprehensive template is expensive. Making the template adaptive without introducing extra complexity is tricky. And ensuring a smoothing transition from prototype to continuous operation in production is difficult. Therefore successful, generic template-based systems are very rare. In practice, template-based approach works better when the same system needs to be repeatedly deployed at different sites with moderate customizations. For example, when a package sorting robot fleet is deployed at different storage spaces, it needs to be customized to adapt to different storage layouts. However, the rest of the workload is pretty much standardized. This is a good use case for a template-based system that formalizes most of the workflow while allowing site-specific configuration to be easily injected.

Akraino is a LE Edge project that aims to integrate multiple open-source projects to supply a holistic edge platform, edge application and developer API ecosystem. Essentially, it's a collection of *blueprints* (or templates) for different use cases, such as connected vehicles, AR/VR applications at edge, 5G MEC systems and time-critical edge compute scenarios. At the time of writing, Akraino is not an actual compute stack implementation, though the team has a long-term goal to define a collection of edge middleware and API that edge programmers can leverage.

Akraino blueprints don't provide abstractions across multiple scenarios. Instead, each of the blueprints is very specific to the described use case. And the quality of blueprint varies greatly. For example, the Connected Vehicle Blueprint for R2 describes some interesting use cases such as accurate location and smarter navigation, however, the blueprint itself doesn't reflect how these scenarios are implemented other than using a generic RPC framework from Tencent (https://wiki.akraino.org/display/AK/CVB+Architecture+Document+for+R2). So, although Akraino mission claims to provide an integrated edge infrastructure, the actual unification is still yet to be seen.

Although we don't have a crystal ball, we can safely predict edge computing will never converge on a single technical stack – besides everything else, Apple is very unlikely to put Android into their devices (Microsoft did but that's a different story). So, this "unified stack" itself is a false proposition. Of course, we don't think anyone who has proposed a unified stack actually believes it, either. The practical goal of these unified stack ideas is to create and grow an ecosystem that hopefully attracts bigger market shares than others.

The edge compute ecosystem will always be segmented. And to create some sort of uniformity and predictability, a solution can't be purely technical. In Chapters 7 and 8, we'll explore the capability-oriented architecture (COA) that offers unification and predictable behavior without converging on a single technical stack.

4.5.3 Multi-access edge computing framework

We've mentioned MEC a couple of times in this book. Given it's a genuine compute framework designed specifically for edge, it seems appropriate to present a proper introduction here.

MEC framework is a simple concept – it enables MEC applications to run on a virtualized infrastructure that is located in or close to the network edge. The goal of MEC is to define a set of standard APIs, application-enabling frameworks, management and orchestration capabilities that allow applications to operate on an edge network on top of physical and virtualized devices. MEC enables cellular operators to open their radio access network (RAN) to third parties. MEC uses cellular networks and 5G as its primary connectivity. And it's designed to be implemented at the cellar base stations and other edge devices.

MEC framework contains system-level components, host-level components and network-level components. MEC applications are hosted on a MEC host. MEC hosts are connected through the network components and managed by the system-level components. Figure 4.6 shows a high-level MEC architecture.

MEC and network functions virtualization (NFV) are complementary concepts in this context. A MEC platform can be deployed as an NFV. And MEC applications can be deployed as NFVs as well.

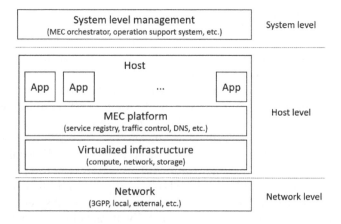

Figure 4.6 MEC architecture.

This concludes our discussion of edge native compute stacks. We will skip typical peer-to-peer systems and distributed systems in general, as they are not constrained to edge computing. Running Kubernetes as an edge compute platform is a popular and important pattern, and we'll dedicate the entire next chapter on the subject.

Next, we'll discuss the holy grail of edge computing – autonomous systems.

4.6 AUTONOMOUS SYSTEMS

My friends and I often discuss how robots will gain self-awareness and take over the world. Although not agreed by most of my friends, I believe that a robot can gain self-awareness by evolving "artificial motivation", which is based on coded instinct, such as an instinct to maintain a continuous power supply. Animals gain such instincts through lengthy evolution. But we can code these instincts into robots to cut the evolution process by a couple of million years, maybe. The interesting thing is, we as human beings have foreseen what could happen, yet we are still trying to make machines around us more intelligent, more self-sufficient and more adaptive. The temptation is clear – if we can get machines to take over routine, danger- ous, error-prone or simply boring tasks, we can make the world a better place for everyone. It's worth the risk, isn't it? Especially, as the creator of the machines, we have a natural confidence that we'll remain in control.

With this little philosophical debate out of the way, we'll switch focus to examine various aspects of an autonomous system. At this point, we are not concerned with self-aware machines. Instead, we define an autonomous system simply as a "self-adaptive automated system". Take a self-driving car as an example – create a controller that automatically turns a car wheel left and right is easy. The tricky part is to make the system automatically adaptive to the environment – different roads, different weather conditions, different traffic patterns and different kinds of animals that leap out from nowhere. The self-adaptiveness needs the machine to understand the physi- cal context, which is punishingly difficult in an open environment such as streets for auto-driving.

4.6.1 From automatic to autonomous

Automatic systems are extremely efficient and often very rigid. They are very efficient in handling the specific tasks they are assigned to, such as punching holes through metal sheets, placing caps on bottles and twisting wires to certain angles. Any updates in the task profile need manual recon- figuration and calibration of the system.

CNC machines, or computer numerical control machines, are broadly used in automated production lines in manufacturing. These machines traditionally handle materials in a subtractive fashion, in which they carve raw materials (by drilling, milling and turning) into usable parts with high efficiency and accuracy. The parts to be handled by CNCs are usually modeled by a computer-aided design (CAD) system so that they can be processed by CNC machines who often operate in cylindrical space. To process a new part, the system needs to reconfigure with a new model file and recalibrated.

An additive manufacturing process such as 3D print creates parts by accumulating raw materials into designed shapes. It's a more flexible and approachable process, and it significantly reduces material wastes. However, in principle it's still a rigid automation system as it doesn't self-adapt to task variations and everything still needs to be precisely planned beforehand by human users.

An autonomous system, on the other hand, is adaptive to task variations and take corrective actions. For example, an automatic integrated circuit (IC) chips packer can line up IC chips that have been dropped on a conveyor belt at random angles into well-formed rows to further packaging, as illustrated in Figure 4.7.

This system is adaptive to various angles of chips because it uses a computer vision system to recognize how the chips are placed, and rapidly generates a new action plan for the robot arms to pick up the chip, rotate and translate it by the appropriate amount, and place it in a well-formed row of chips. This rapid feedback and reactive loop enables the system to adjust to variations in work profiles, hence laying the foundation of an autonomous system.

4.6.2 The human factor

Humans can be forced to endure lengthy and repetitive work – such as packaging playing card decks into boxes. An experienced and committed worker can package hundreds of decks in one shift. Such manual labor is unlikely to completely go away in years to come, because the cost of

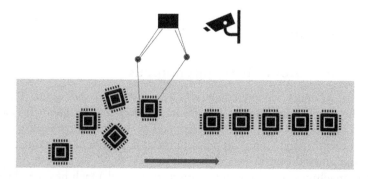

Figure 4.7 IC chip sorting system.

automating such systems often overweight (with a huge margin) the cost of manual labor. Furthermore, packaging playing card is a low-risk task, which lowers the motivation and necessity to use more expensive machines.

Humans are also adaptive. When given a repetitive task, people often can figure out smart ways to perform the task more efficiently. If you observe how a worker package a deck of playing card, you'll find that any unnecessary movements have been eliminated. Tens of actions are performed with great precision to cut the deck out of a continuous deck of cards, fold the box, place the cards and close the box. Some of these experiences can be codified into automated systems. And some of them can be taught to machines through methods such as reinforced learning. To a great extent, the goal of many autonomous systems is to be able to efficiently mimic what humans do, and do it in an enduring, consistent way.

An important benefit of an autonomous system is to eliminate human errors. Humans make mistakes. This is just how we operate, because throughout the long path of evolution, we favor adaptivity over reliability. This makes us unsuitable for long, repetitive tasks. In some other cases (such as handling dangerous chemicals), the cost of human errors is very high, so we have to search for alternative solutions. Furthermore, some modern tasks such as soldering IC chips are so delicate and complex that they can't be reliably performed by humans at scale. Autonomous systems learn from human experiences so that they can handle different task variations and then remove humans from the workflow.

4.6.3 Autonomous system architecture

Autonomous systems, especially intelligent autonomous systems, are often studied as intelligent autonomous agents (IAAs). An agent is assigned with goals. It perceives the environment and generates planned actions based on policies, local states and environmental factors to achieve its goals. Figure 4.8 illustrates the architecture of a simple autonomous system. The system collects data from environment. The data is saved to a local state store. And events are sent to the controller through a local message bus. The controller makes decisions based on the current state and received events

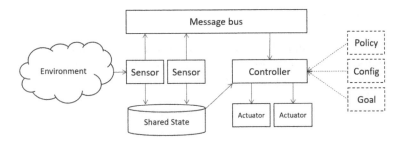

Figure 4.8 Simple autonomous system architecture.

and drives the actuators to react. In addition, an administrator can inject policies, goals and configurations through a (remote or local) management interface.

The controller is the core of the system. It can be a simple reactive controller that reacts to events, or a template-driven controller that runs assigned *recipes* for matching conditions, or a fuzz logic controller (FLC) that use fuzzy logic algorithms to make decisions, or an AI-based controller that user sensor readings as inputs to a AI model to generate necessary actions.

A more complex autonomous system may be comprised with multiple intelligent agents. These agents communicate with each other and collectively generate activities to achieve goals of the system. There are multiple ways to orchestrate these agents: a rule-based orchestrator, self-coordination using a reactive programming pattern or a complex system in which global behaviors emerge while individual agents seek to optimize their own sub-goals.

4.6.4 Cloud-assisted robotics

Robotics is one of the most representative cases of autonomous systems. At the time of writing, the industry mostly focuses on industrial robotics and self-driving cars. However, it's easy to anticipate everyday robots and more autonomous agents will become the norm in the years to come. Because this is a book about software architecture, we won't spend much space on the hardware aspects of robotics. Instead, we offer a list of cloud services that are designed to accelerate robotics developments:

- Fleet management

Cloud offers scalable management of scattered robots. This includes applying consistent configurations and policies, performing bulk software updates, monitoring the entire fleet through a single pane of glass, coordinating actions of multiple robots (such as resolve conflicts in planned paths) and other central management functionalities.

- Simulation

Reinforced learning and genetic algorithms are often used to train intelligent autonomous agents. Both methods require a huge number of iterations, making it infeasible to perform such training in real environments. High-fidelity simulations allow developers to emulate real physical environments for such training. For example, Microsoft AirSim (Aerial Informatics and Robotics Simulation) uses Unreal game engine to simulate physical environments such as buildings and streets for training self-driving cars and intelligent drones.

- AI training

In general, cloud can offer huge amount of compute resources (both CPU and GPU) for paralleled AI model training. Furthermore, because many IoT devices are pumping data to the cloud, cloud can directly tap into the data sources to perform training.

- Online development environment

Many cloud platforms are working on providing an end-to-end device life cycle management and application development experience on cloud. When it comes to developer experience, Microsoft Azure is in the lead. It has online editing experience, complete DevOps pipeline, integrated source control, AI model authoring tool as well as device management, digital twin, simulation and other related features, making it an ideal choice for online robotics development.

- Connectivity

Cloud platforms offer private links between robotics and cloud to provide secured channel to protect data in transition. Some platforms also provide satellite-based links (such as Azure Orbital, SpaceX Starlink and Amazon's Ground Station) to provide connectives in remote areas.

4.7 SUMMARY

This chapter reviews a few groups of design patterns that push compute from cloud to edge, including edge acceleration patterns, cloudlet patterns, edge function patterns, cloud compute stack on edge, edge native compute stack and autonomous systems.

Edge computing is about compute in contexts. Bringing the cloud power and efficiency into the real-world contexts enables powerful edge compute scenarios. At the time of writing, the landscape of edge computing is heavily segmented. From a software perspective, some sort of uniformity will improve developer productivity by making their code adaptable to different variations of dynamic edge environment. And this will be the focus of the second half of the book. We'll first examine how Kubernetes offers some infrastructural support for a consistent application model. Then, we'll spend time on COA that aims to provide a unified experience on edge.

Chapter 5

Kubernetes on edge

The Borg is an alien group in the science fiction television series *Star Trek*. They are cyborgs connected to a *hive mind* named "the Collective". Their ultimate goal is to achieve perfection. And they do this by injecting molecular machines into various alien species to transform the victims to controlled *drones* that contribute technology and knowledge to the Collective. The collective mind is capable to heal damaged Borgs. It also generates tactics for the Borg to deal with threatening situations.

Around *Stardate* 92016 (or mid-2014), Google announced an open-sourced project named Kubernetes, which had the root of an internal Google project, Borg. Kubernetes (often written as k8s) works somewhat like how Borg the alien group works – it installs agents to compute nodes to link these nodes into a central control plane that monitors the health of nodes and schedules workloads on the compute nodes. When a node fails, the scheduler shifts its workloads to other healthy nodes. And the control plane can be configured with scalers that adjust number of compute instances when workloads change. This design allows a group of compute nodes to be organized into a compute plan that offers scalable and highly available hosting environments for applications.

Kubernetes was originally designed for servers and datacenters. As we've discussed in previous chapters, a common pattern of edge computing is to push cloud infrastructure design into edge. The same force pulls down Kubernetes to the edge as well. A side effect of this movement is that Kubernetes becomes pervasive across cloud and edge, providing a consistent infrastructure management and application deployment experience for ubiquitous computing.

Given the significant market share of Kubernetes, we dedicate this chapter to discussing how Kubernetes works on edge, and how you can design your solutions leveraging Kubernetes on edge.

This chapter is not an introduction to Kubernetes. Although we'll briefly explain Kubernetes concepts as we encounter them, we assume you are already familiar with Kubernetes concepts and know the basics of how Kubernetes work.

5.1 AN ANATOMY OF A KUBERNETES CLUSTER

A Kubernetes cluster consists of one or multiple *nodes* that host Kubernetes *control plane* as well as user *workloads*. The control plane exposes an *application programming interface (API) server* that worker node agent, which is called *kubelet*, connects to report health and to get workload allocations. Kubelet schedules *pods*, which are groups of containers running the actual workloads. User connectivity to the pods is provided by a *kube-proxy*, which also runs on every worker node.

The main component of the control plane is the *kube-apiserver*, which exposes Kubernetes control plane API. It uses *etcd* as its backing store. The actual component that is responsible for pod scheduling is called *kube-scheduler*. It watches for new pods without assigned nodes and chooses where to place these pods. There are also other control plane components, such as *kube-controller-manager* and *cloud-controller-manager*, but they are out of the scope of this book. Please consult Kubernetes online documents for more details.

On a node, the *kubelet* works with different container runtimes (such as *dockerd*) through a common Container Runtime Interface (CRI). A container runtime handles container-related tasks such as creating containers from container images. The Open Container Initiative (OCI – see https://opencontainers.org/) defines common specs for container runtimes and container image formats. A container runtime following the OCI specs is called an OCI-compliant runtime. Figure 5.1 illustrates the structure of a Kubernetes node. In addition to Kubelet, OCI and CRI, the diagram also shows some other common extension points, including pluggable software-defined network (SDN) through Container Network Interface (CNI), pluggable device types (such as a graphics processing unit [GPU]

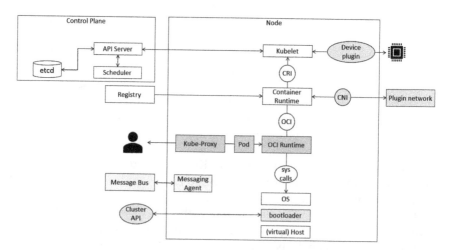

Figure 5.1 Structure of a Kubernetes node.

and field-programmable gate array [FPGA]) through device plugin, and cluster provisioning operations through Cluster API.

All these interfaces and extension points provide us opportunities to extend Kubernetes in various ways to meet your project needs. The following sections provide more details on these extension points, and we'll provide examples of how these extension points are leveraged later in this chapter.

5.1.1 Kubelet

Kubelet is the main node agent running on each node. It registers the node with the control plane. Then, it takes pod description, which is called a *PodSpec*, from the control plane and ensures the containers described in the PodSpec are running and healthy.

A Kubelet communicates with the control plane API server through the API server's endpoint. It also hosts a Hypertext Transfer Protocol Secure (HTTPS) endpoint itself for the API server to callback to it. The API server uses the endpoint to fetch pod logs, to perform port forwarding and to launch pod directly on the node. At the time of writing, the API server doesn't verify the Kubelet's serving certificate, which means the connection is subject to *man-in-the-middle attack*. Hence, when you try to run control plan and worker nodes over untrusted networks – such as running work nodes on edge while keeping the control plan on cloud, you'll need to secure the connection yourself.

5.1.2 CRI

Kubelet relies on a *container runtime* that handles actual container provision and management. In earlier days, kubelet supported a couple of container runtimes (e.g., dockerd) through an internal interface. This meant to support a new container runtime, the upstream kubelet code needed to be updated. CRI was created to address this issue. It consists a specification, a *protobuf* API and libraries for a container runtime to integrate with kubelet. Since its original alpha release, several CRI runtimes have been in the work, noticeably CRI-O(https://github.com/cri-o/cri-o) and frakti (https://github.com/kubernetes/frakti).

The CRI API defines necessary methods for container manipulation, such as creating, starting, stopping and removing containers. It also defines several other methods such as attaching to a running container and setting up port forwarding. The following code snippet shows some of the container methods defined by the API:

```
rpc CreateContainer(CreateContainerRequest) returns
(CreateContainerResponse) {}
rpc StartContainer(StartContainerRequest) returns
(StartContainerResponse) {}
rpc StopContainer(StopContainerRequest) returns
(StopContainerResponse) {}
```

```
rpc RemoveContainer(RemoveContainerRequest) returns
(RemoveContainerResponse) {}
rpc ListContainers(ListContainersRequest) returns
(ListContainersResponse) {}
rpc ContainerStatus(ContainerStatusRequest) returns
(ContainerStatusResponse) {}
rpc ExecSync(ExecSyncRequest) returns (ExecSyncResponse) {}
rpc Exec(ExecRequest) returns (ExecResponse) {}
rpc Attach(AttachRequest) returns (AttachResponse) {}
rpc PortForward(PortForwardRequest) returns
(PortForwardResponse) {}
```

CRI allows different container runtimes to be plugged into Kubernetes. And in theory, you can plug in a CRI that is not based on any container engines. For example, you can author a CRI that manages loose processes and even threads that optionally use your own isolation layer instead of relying on existing container engines.

5.1.3 OCI

Open Container Initiative (OCI) is a set of open specifications for image formats (defined by a *image-spec*) and container runtimes (defined by a *runtime-spec*). The OCI image format defines how application files can be packaged in an image bundle. And the OCI runtime specification outlines how the bundle is unpacked and mounted on a file system.

Docker had donated its container format and its runtime, *runC,* to the project to serve as the foundation. And the industry has contributed a few other OCI compatible runtimes such as a low-memory runtime *crun* (https://github.com/containers/crun) written in C and a hypervisor-based runtime *runV* (https://github.com/hyperhq/runv).

5.1.4 Device plugin

By default, kubelet manages only central processing unit (CPU) and memory resources. Kubernetes has a *device plugin framework* that allows hardware vendors to register other hardware resources, such as GPUs, high-performance Network Interface Card (NICs), InfiniBand, storage devices and FPGAs to the control plane (through kubelet). The plugin registers these devices with kubelet and monitors health of these devices.

Then, a container can request these hardware resources in its container specification. For example, the following container specification requires one NVIDIA GPU as a *resource limit.* When the container is scheduled, the control plane ensures the resource limit is satisfied by placing the container to the node (kubelet) where the GPU has been registered:

```
apiVersion: v1
kind: Pod
```

```
metadata:
  name: cuda-inference
spec:
  restartPolicy: OnFailure
  containers:
  - name: cuda-inference
    image: "my-container:latest"
    resources:
    limits:
      nvidia.com/gpu: 1
```

Device plugins allow you to model hardware requirements into your application. For example, if your application container needs a camera from a specific vendor, you can define a camera device plugin to register your device with the physical node to which the camera is attached. Then, when your application is scheduled, the container will be placed on the correct server because the schedule tries to satisfy the resource limit (as shown in the previous example) specified in your container specification.

Devices, such as sensors, controllers and cameras are considered leaf devices in edge scenarios and while device plugins offer some help, they still present a management challenge. This is where projects like Akri (https://github.com/deislabs/akri) can help as they expose leaf devices as first-class resources in a K8s cluster. Akri is extending the device plugin framework to make it work for edge scenarios.

5.1.5 CNI

CNI (Container Network Interface) is not a Kubernetes-specific concept. It consists of a specification and associated libraries for writing network interface plugins for Linux containers. CNI plugins are shipped as executables that are invoked by container management systems such as Kubernetes. They implement an ADD method and a DELETE method (along with CHECK and VERSION) to add or remove a container from a network.

Kubernetes defines a flat network model: every pod has its own IP address; Pods on any nodes can communicate with each other without NAT. Kubernetes uses two types of CNI plugins to interact with the underlying network – CNI network plugins, which we've introduced above, and CNI IPAM plugins. CNI IPAM plugs are responsible for allocating and releasing IP addresses for pods.

When you deploy Kubernetes-based solutions to either cloud or edge, you can choose from existing CNI plugins to match your network environment. Or you can implement your own CNI plugins as needed. For example, you may want to create a Z-Wave plugin for your smart home component to communicate with each other when they are managed by a lightweight Kubernetes cluster deployed on edge (such as on a home gateway).

5.1.6 Cluster API

Cluster API concerns with cluster creation, configuration and management. Cluster API itself runs on a Kubernetes cluster, which is referred as a *management cluster*. Then, you can use Cluster API (hosted on the management cluster) to provision and manage other *workload clusters*. And it's a common practice to create a local, temporary *bootstrap cluster* to bring up the management cluster.

Where does the bootstrap cluster come, you ask? Good question! Eventually you'll need to use some tools like *kind* (https://kind.sigs.k8s. io/), kubespray (https://kubespray.io/), Kubeadm (see https://kubernetesio) or kops (https://github.com/kubernetes/kops) to get the ball rolling.

Let's say eventually you have a Cluster API running somewhere. Now, you have a control plane for control planes. This is useful especially when you need to manage a fleet of Kubernetes clusters for your different edge scenarios. You can declaratively define your clusters and use Cluster API to automate the cluster provisioning and management processes.

A full-scale, production-ready Kubernetes cluster needs quite a bit of resources to run. A typical *etcd* cluster itself needs 2–4 CPU cores, 8G of memory and 1Gb network connection to run well. On top of that, we need multiple nodes and compute resources to support multiple instances of the API server and other control plane components. It's not always practical to push such a heavy infrastructure to the edge. Fortunately, with the aforementioned extension points, we can apply some innovative designs to split the cluster up and deploys only the minimum required infrastructure to the edge to support the workload. In the next sections, we'll introduce some of the existing solutions that use lightweight clusters, custom kubelet implementations, lightweight CRI and container runtimes.

5.2 LIGHTWEIGHT KUBERNETES CLUSTERS

We've observed two methods to make a Kubernetes cluster fit to edge: to slim down the whole cluster or to split the control plane and compute plane so that only the compute plane is deployed to the edge compute resources. Obviously, slimming down the whole cluster takes a lot of efforts – therefore there's only a couple out there (at the time of writing). Splitting the cluster is slightly more approachable, and we'll present some of the projects in this section.

5.2.1 Minikube

Minikube (https://minikube.sigs.k8s.io/) is a tool that allows you to run a single-node Kubernetes cluster on your development or test environments (such as your desktop). It's a quick and easy way to get a local Kubernetes going for dev-test purposes. It's not designed for production workloads, but more of a learning tool to get something going quickly. As we don't see Minikube being applied to production edge workloads, we'll not go into more detail in this book.

5.2.2 K3s

K3s (https://k3s.io/) is a certified Kubernetes distro made by Rancher. It's shipped as a single 40MB binary with less than 250MB memory consumption at runtime (for one control plane node and a single agent). It combines the entire Kubernetes control plane, kubelet as well as container runtime (*containerd*) in a single process, which is launched as a *systemd* or *openrc* service. In addition to *etcd*, it supports other state stores such as SQLite, Postgres and MYSQL, and it also supports an in-process state store called *dqlite*.

K3s reduces the size of Kubernetes by going into the upstream source code and removing about 3 million lines of code from the code base. Features like in-tree cloud providers, in-tree storage drivers, alpha features, legacy and non-default features are removed along with Docker-related code. At the same time, it also adds some features to support common scenarios including Transport Layer Security (TLS) management, Helm chart support, CoreDNS, Traefik and Flannel supports. At the time of writing, K3s uses the Docker container runtime. It has plans to switch the runtime to k3c (https://github.com/rancher/k3c), a container runtime created to support lightweight Kubernetes scenarios.

K3s is certainly much more lightweight than K8s. And it works well with popular ARM architecture CPUs. So, it's ideal to be deployed on edge gateways or small edge servers (such as Raspberry Pis). You can also create a highly available K3s deployment with multiple K3s server nodes (with an externalized state store) joined behind a load balancer to serve the Kubernetes API.

5.2.3 MicroK8s

MicroK8s (https://microk8s.io/) is a single-package Kubernetes distro made by Canonical. Its primary target audience is developers who want to quickly set up a single-node Kubernetes clusters for development and CI/CD purposes. At the same time, it also claims to be applicable to edge and Internet of Things (IoT) scenarios with dynamic node management for single or multi-node clusters. According to its documentation, it's the smallest certified K8s distribution for production workloads. It has an addon model that allows different extensions to be deployed, including Istio, Knative, Helm and Kubeflow. MicroK8s runs natively on Linux systems with *snap* (a software deployment and package management system developed by Canonical); and it runs on Windows and macOS through *multipass* (a VM management tool also developed by Canonical).

The minimum requirement of running MicroK8s is 2GB of RAM (4GB is recommended), and about 300MB of disk space (20GB is recommended).

So, it will be tight to run applications on Raspberry Pi 3; but Raspberry Pi 4 has more memory to support the extra loads.

Canonical's snap package manager supports three *confinement levels* that isolate applications from the system: *strict*, *classic* and *devmode*. MicroK8s uses the strict confinement level, which means it cannot access files, networks, processes or any other system resources without requesting specific access via an interface. This is great for edge scenarios as you can protect the hosts from broken or vouge applications deployed by the orchestrator.

Having a slimmed-down version of Kubernetes on edge is a noble idea, but even those trimmed down K8s distros are far away from being able to run on real low-capability devices. There are some options like custom kubelets as we will see later in this chapter. Another aspect to consider is manageability concerns when we consider large-scale edge compute scenarios with many scattered clusters. How do you deploy application to multiple clusters? How do you maintain visibility across them? And how do you load-balance workloads across cluster boundaries? Later in this chapter, we'll introduce a few federated solutions that are specifically designed to answer these questions. At the moment, we'll spend the next section on a different approach – to split the cluster to keep the control plane centrally managed on cloud, and to keep only the compute plane on edge.

5.3 SEPARATING THE CONTROL PLANE AND THE COMPUTE PLANE

Having a centralized control plane over scattered compute resources is almost always desirable: it gives you better visibility; it gives you more streamlined management experience and it provides better security around your control plane. However, when we try to split a Kubernetes cluster up, we'll need to answer a different set of questions: how do we secure the communication channel between the control plane and the compute plane? Would occasional network drops mess up heartbeat signals that lead the control plane to mistakenly believe a worker node has failed? What happens to the worker nodes when they are detached from the control plane? This section presents a couple of solutions that try to address these issues.

5.3.1 KubeEdge

KubeEdge (https://kubeedge.io/) is a Cloud Native Computing Foundation (CNCF) project. It's a combination of two things: a lightweight Kubernetes implementation on edge (it uses a custom kubelet, which we'll cover later in Section 5.4 of this chapter) and a system to establish and manage data pipelines between cloud and edge. We won't go into more detail on the data pipelines in this chapter. Instead, we'll discuss how KubeEdge separates the control plane and the compute plane, and how it enables the compute plane to continue to operate when it's temporarily disconnected from the control plane.

The control plane makes scheduling decisions. These decisions are written to a metadata store for the node agents to pick up and carry out. So, if we make the decision records available to the node agents, the node agents can operate autonomously even when they are disconnected from the control plane. This is the approach KubeEdge takes – it replicates the decision records to the edge so that the node agents operate based on these records regardless of if they are connected to the control plane or not. Data synchronization between cloud and edge is done by a *cloud hub-edge hub* pair (for a complete view of the architecture please see https://github.com/ kubeedge/kubeedge/blob/master/docs/images/kubeedge_arch.png).

This design allows the control plane to push scheduling decisions to the edge when connected. And when the edge is disconnected from the control plane, it can use the cached results to perform certain recovery actions. For example, when a failed node comes back, it can read its prior workload allocation for the local state cache and rebuild its states.

5.3.2 OpenYurt

OpenYurt (https://github.com/alibaba/openyurt) is an open-source project led by Alibaba, who announced the project just a few weeks before this text is written. The project aims to provide continuous Kubernetes node operation on edge when connection to the control plane is interrupted. Architecturally, OpenYurt looks very similar to KubeEdge (based on the information we can collect so far). It introduces an *edge-hub* that serves as a reverse proxy for kubelete. The proxy persists and serves kubelet state data from its cache when the network is disconnected. This allows kubelet to continue its operations when disconnected from the control plane. When the edge nodes are disconnected, server-side operations, such as scheduling decisions are cached in server-side *Yurt controllers*. And data is synced to the edge when the network connection is restored.

Open KubeEdge and OpenYurt achieve separation of the control plane and the compute plane by data replication. The upside of this approach is that minimum changes are required. The downside, however, that the edge nodes can't make any scheduling decisions by themselves. So, both systems are suitable for mostly connected systems. When network outage persists, quality of services on the edge may degrade and cause prolonged outage. For example, a service may have been deployed to two different edge nodes for high availability. If both nodes fail, the service is broken even if there are other healthy nodes available, as no one is making new scheduling decisions to leverage the remaining nodes.

In Chapter 7, we'll introduce a *quicksilver* scheduling system that allows workloads to reschedule themselves without a centralized control plane.

KubeEdge and OpenYurt split Kubernetes clusters by introducing additional layers between the control plane and kubelet. In the next section, we'll examine a couple of solutions that goes a step further.

5.4 CUSTOM KUBELET IMPLEMENTATIONS

A kubelet represents a compute node to which the scheduler can allocate workloads. When kubelet was designed, it was meant to work on a fully connected cluster by statically representing a compute node. What if we need to manage a dynamic compute plane that is composed of remote, scattered resources such as container hosts managed by a cloud platform, cloud services or low-power devices that can't support kubelets by themselves? This section introduces two custom kubelet implementations that extend the traditional compute plane to a much broader range.

5.4.1 Virtual kubelet

Virtual kubelet (https://github.com/virtual-kubelet/virtual-kubelet) is an open-source implementation of Kubernetes-kubelet interface. The idea is to be able to represent arbitrary compute resources as a Kubernetes node so that Kubernetes can schedule workloads on them. For example, there are *providers* that enable various cloud-hosted compute resources, including Azure Container Instances and Amazon Web Services Fargate, to be connected to a Kubernetes control plane. Although there are no official edge-based providers at the time of writing, conceptually someone can create an edge-based provider that represents far-edge or extreme-edge compute resources for edge scenarios.

So, virtual kubelet "pretends" to be a physical compute node by implementing kubelet interface methods – such as *CreatePod()*, *UpdatePod()* and *GetPod()*. Then, it turns around and uses vendor-specific APIs to manage underlying compute resources.

In early 2020 (which still sounds like distant future, as depicted by numerous science fiction novels and movies), the authors of the book implemented a virtual kubelet provider for Azure Sphere. Azure Sphere (https://azure.microsoft.com/en-in/services/azure-sphere/) is a secured IoT solution released by Microsoft. It includes hardware, OS as well as cloud components for device management. It has its own application package format and its own application distribution system that can push new bits of applications to devices through over-the-air (OTA) updates. The first generation of the devices has 4MB of onboard memory, which means it's impossible to run Kubelets on the devices themselves. The idea of the virtual Kubelet provider is to deploy virtual Kubelet instances on a field gateway computer. And the virtual Kubelet instances work with the Sphere devices through Azure Sphere standard API and tool. Furthermore, we wrap Azure Sphere application packages in standard Docker images as a separate layer on top of an empty *from-scratch* base image. This allows Azure Sphere applications to be packaged and shared through standard Docker tools and services such as Docker CLI and Docker registry. When a pod is scheduled, the virtual Kubelet extracts the application package out of the container image and uses Azure Sphere APIs to schedule the application onto attached Azure Sphere devices.

The implementation allows Azure Sphere devices to join into a bigger IoT solution that involves off-device parts such as edge servers and cloud services. In such cases, a user can create a single Kubernetes deployment profile that describes all components. And when the profile is applied, Azure Sphere devices are updated along with other cloud and edge resources as an integrated solution.

Just for the record, Figure 5.2 shows a system with two Azure Spheres jointly managing a "complex" IoT device with a LED light and a LED counter. You can see how the application is scheduled as two pods running on two Azure Sphere devices (within one deployment).

When you have mixture of devices on a Kubernetes cluster, you'll probably need to set up *node affinities* so that pods are scheduled only to compatible nodes – you certainly don't want to schedule a Redis container onto an Azure Sphere device, for instance. Kubernetes allows you to define both node affinity, which "attracts" pods to nodes, and anti-affinity (called *taints*) that "repels" pods from nodes. Furthermore, you can attach *tolerance* to pods so that they can be deployed to nodes with matching taints. In our case, we defined a "sphere" taint on the Azure Sphere node and associated the same "sphere" tolerance to the pods that we want to schedule onto the Azure Sphere devices.

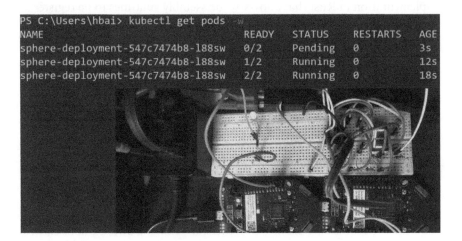

Figure 5.2 Virtual kubelet with Azure Spheres.

5.4.2 Krustlet

Before we introduce Krustlet, we need to spend some time on WebAssembly, just in case you are not familiar with it yet. WebAssembly (or *Wasm* for short) is an instruction format for a stack-based virtual machine. You can compile programs written in high-level programming languages, such as C++ and Rust, into WebAssembly binary format and run them natively on modern browsers (such as Chrome, Firefox and Edge). WebAssembly code executes generally faster than JavaScript code, making it ideal choice to perform demanding computing tasks to provider a faster, more fluent user experience. It also allows some legacy (especially C++-based) applications and libraries to be cross-compiled and run natively in browser for highly interactive scenarios such as online gaming, animation and simulations.

Because of the pervasive reach of browsers, WebAssembly has the potential to be a massive edge compute plane that can touch literally all personal computers out there. At the same time, running WebAssembly outside the browser also gets increasing attention. For instance, a WebAssembly System Interface (WASI, see https://wasi.dev/) has been proposed to provide a set of operating-system-like features, such as file accesses, to WebAssembly applications. This allows legacy code with system calls to be cross-compiled into the portable WebAssembly binary format and runs anywhere WASI is supported, such as standalone WebAssembly runtimes including *wasmtime* (https://github.com/bytecodealliance/wasmtime), WebAssembly Micro Runtime (WAMR, see https://github.com/bytecodealliance/wasm-micro-runtime) and *wasm3* (https://github.com/wasm3/wasm3).

We'll return to the idea of a "browser-based cloud" again in Chapter 8. Now, let's go back to Krustlet.

Krustlet (https://deislabs.io/posts/introducing-krustlet/), or Kubernetes-rust-kubelet, was announced by Deis Labs in early 2020. It's a Kubelet implementation in Rust that allows WebAssembly runtimes to be registered as Kubernetes nodes so that the control plane can schedule containerized WebAssembly modules to these runtimes.

Both Virtual Kubelet and Krustlet use custom Kubelet implementations to adopt standard pod operations into platform-specific instructions. And when the Kubelet implementation is kept off device (on a field gateway, for instance), the physical devices that host the actual workloads need only the required runtimes (such as Azure Sphere runtime and WebAssembly), reducing the resource demands on devices to bare minimum. This makes it possible to use a centralized control plane to manage lower-power devices.

However, neither solution is designed for massive device deployments. Because they still use the kubelet interface and register each device as an individual node, their scalability is subject to Kubernete's limitations. At the time of writing, Kubernetes control plane tends to struggle when the cluster size exceeds 500. Although some have claimed to successfully running a 5,000-node cluster, a considerable amount of extra care is needed to maintain such a big cluster.

To achieve massive-scale deployment, we'll need something different, such as cascaded scheduling, or distributed scheduling that we'll discuss in Chapter 7.

In the next section, we'll go yet another step deeper to examine what can be done with lightweight container runtimes and introduce a few noticeable projects.

5.5 LIGHTWEIGHT CONTAINER RUNTIMES

Informally, there are two types of container runtimes – a low-level container runtime that concerns with just container CRUD operations and a high-level container runtime that offers additional features such as images management and network management (often through CNI). Roughly, you can consider runtimes that implement CRI as a high-level container runtime, while runtimes that implement OCI as a low-level container runtime.

A few years back, Docker was the dominating container runtime in the industry. Later, the runtime was split into *containerd* (that implements CRI) and *runc* (that implements OCI), which roughly correspond to a high-level runtime and a low-level runtime, respectively.

Just for historical context, Kubernetes natively supported another runtime named *rkt* created by CoreOS who later acquired by RedHat. The project was declared dead in early 2020.

Since then, quite a few container runtimes have been created for different environments and scenarios. They can be categorized into three broad types – isolation-based container runtimes that use Linux kernel primitives such as *cgroups* and *namspaces* to provide isolated environments (but not a sandboxed environment) for hosted processes; hypervisor-based container runtimes that run a dedicated kernel or a minimum kernel (such as *Unikernel*) directly on top of virtual machine hypervisors to provide stronger isolation between containers and bare-metal solutions that use container runtime as the PID 1 process. Figure 5.3 illustrates how components are stacked on container hosts with different container types.

This section follows both the CRI-OCI vector and the isolation vector to present a few custom container runtimes and their designed use cases. Of course, what we cover here is just a subset of all the container runtimes out there, as new ones constantly popping up here and there.

5.5.1 CRI-O

CRI-O (https://cri-o.io/) is a CRI implementation that allows any OCI-compatible runtimes to be plugged into Kubernetes. It offers implementations of common features such as image management, storage and networking and allows OCI-compatible container runtimes to be plugged in through configuration.

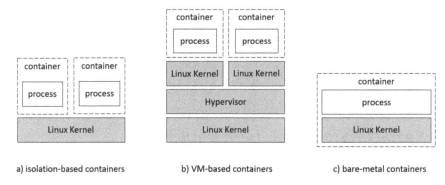

a) isolation-based containers   b) VM-based containers   c) bare-metal containers

Figure 5.3 Comparison of different containers.

CRI-O's value becomes the most apparent when you need to use the same Kubernetes control plane to manage heterogeneous devices that use different low-level container runtimes. In such cases, CRI-O tools can provide unified management experiences across different runtimes.

5.5.2 Kata containers

Figure 5.3 shows that traditionally, containers share the same OS kernel on the same host. If an adversary escapes from the container and invades the kernel, it can do damage to other containers on the same host. VM-based container uses a dedicated OS kernel for each of the containers (as shown in Figure 5.3(b)). This provides a stronger isolation among containers hence is more secure.

Kata containers (https://katacontainers.io/) is an OpenStack project. It is a combination of two earlier projects, namely Intel Clear Containers and Hyper.SH runV. It runs containers in dedicated, lightweight virtual machines and provides an OCI-compatible runtime implementation as well as a Docker shim to connect to Docker tools.

Kata supports different hypervisors such as QEMU, NEMU and Amazon Firecracker. It runs an agent in the virtual machine, and the agent is controlled by its runtime (as well as the Docker shim) through a remote gRPC interface. Figure 5.4 illustrates how Kata runtime can work side-by-side with other container runtimes on the same Kubernetes cluster:

5.5.3 Windows containers

Windows containers run natively on Windows servers (with container enabled). It provides both process isolation and hypervisor-based (Hyper-V in this case) isolation. For example, the following Docker command launches a new container using the Hyper-V isolation using the --isolation switch:

```
docker run -it --isolation=hyperv mcr.microsoft.com/windows/
servercore cmd
```

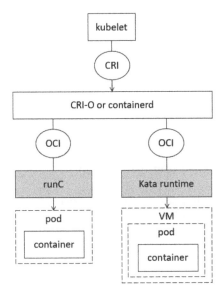

Figure 5.4 Kata runtime and runC runtime on the same cluster.

5.5.4 Nabla containers

Nabla containers (https://nabla-containers.github.io) is made by IBM. It's not a VM-based container, but it does use unikernel to reduce the attack surface between the container and the host. Specifically, only seven system calls (read, write, exit_group, clock_gettime, ppoll, pwrite64 and pread64) are allowed. All other calls are blocked through a Linux *seccomp* policy.

5.5.5 k3 container solution

In addition to the k3s project mentioned above, Rancher has a grand plan to build up an end-to-end *k3* container solution that consists of the lightweight Kubernetes distro *k3s*, a lightweight container runtime *k3c*, a lightweight, container-based operation system *k3os* and a k3s cluster bootstrapping tool *k3d*.

An end-to-end, secured, lightweight Kubernetes stack on top of bare metal is very attractive to edge scenarios for a couple of obvious reasons. First, a single vendor reduces the probability of incompatibility problems among components from different vendors. Second, the end-to-end solution removes unnecessary overhead and makes effective use of on-device resources. Of course, the global edge compute remains heterogeneous for sure. But within the scope of a specific project, a unified, end-to-end solution is often preferable. At the time of writing, the k3s container solution seems to be comprehensive and promising, though some of its components are still in the experimental phase.

5.5.6 Other noticeable container runtimes

Finally, we list out some other container runtimes that have emerged:

- **crun** (https://github.com/containers/crun)

A low-memory footprint OCI-compatible container runtime written in C. According to its documents, it runs nearly 50% faster comparing to *runc* when tested to sequentially run 100 containers.

- **runV** (https://github.com/hyperhq/runv)

A hypervisor-based, OCI-compatible container runtime. It supports different hypervisors including KVM and Xen.

- **Frakti** (https://github.com/kubernetes/frakti)

A CRI implementation. It uses a hypervisor-agnostic container runtime *hyperd* (https://github.com/hyperhq/hyperd) that serves as an API wrapper around *runV*. It's designed to work with Kata containers.

- **gVisor** (https://github.com/google/gvisor)

A sandbox for containers. It implements a user space kernel (Sentry) that implements system call interfaces. It has a *runsc* OCI-compatible runtime that launches containers in its sandboxed environment. And it uses a *Gofer* process to isolate file/network IO operations. This provides a stronger isolation of containers. We'll discuss container isolation more in Section 5.7.1.

- **Singularity** (https://sylabs.io/)

We are including Singularity here because it seems to be quite popular in academic circles. It's not a lightweight container runtime. Instead, it is a container runtime designed to encapsulate and run HPC workloads. It used its own container image format but later adopted OCI in later versions. At the time of writing, there's also a project to create a CRI interface.

- **Podman** (https://podman.io/)

Some scenarios, for example banking or automotive, are highly regulated and require a high degree of security. In order to take advantage of density and small footprint container runtimes provide, as opposed to virtual machines, we need to make sure that the container runtime does not have root access, which is often referred to as "rootless containers". While there are other runtimes such as rootless Docker, Podman has gained a lot of momentum. Podman is OCI compliant and allows you to run containers in the context of a non-privileged user.

The paragraph of lightweight container runtimes wouldn't be complete in the context of edge scenarios if we did not address one of the hardest requirements for low-capability edges, which is supporting real-time scenarios.

In a broader sense to be real time means that a system or service needs to operate under given timing constraints. You need to decide between hard real time and soft real time, in which hard real time means that a result needs to be provided before a given deadline arrives, whereas soft real time often refers to the best effort. In other words, missing hard real-time deadlines can have dramatic consequences, just think about braking systems in a car, while missing a soft real-time deadline might only cause some degradation in service quality

What does this have to do with container runtimes and hypervisors you may ask? Container runtimes and hypervisors are usually not built with real-time support in mind, meaning they usually do not provide deadline guaranties for startup failover, etc.

The good news is that there are solutions for this as well called real-time hypervisors or runtimes. L4Re (https://l4re.org/overview.html) is a popular example that finds use in many edge real-time scenarios.

In the past few sections, we've examined different approaches to fit a Kubernetes cluster on to edge devices. Like mentioned earlier, because these approaches use Kubernetes cluster-level APIs and extension points, they are subjected to scalability limitations of a single Kubernetes cluster. To manage devices at a large scale, we need to think of ways to manage devices across multiple Kubernetes clusters, which is the topic of the next section.

5.6 CLUSTER FEDERATION

We often need to consider federated clusters not only to overcome scaling limitations of Kubernetes itself, but also to enable additional scenarios such as geographically distributed deployments and hierarchical deployment topologies that can't be easily expressed with a single Kubernetes cluster. In this section, we'll first introduce some typical federation topologies. Then, we'll discuss some of the key challenges in managing a federated environment.

5.6.1 Federation topologies

We've observed several possible federation topologies – fan out, primary-secondary, hierarchical, multi-site span and hub-and-spoke. These topologies are often used in large-scale deployments, especially geographically distributed deployments.

- **Fan out**

This is a simple scale-out topology. In this topology, application packages are shipped to distributed clusters and deployed independently. There's no cross-cluster communication among application instances. And a global traffic manager routes user traffic to different clusters based on different policies such as by measured average latency or proximity. A federated control plane is not mandatory in this case, because there are no cross-cluster scheduling decisions to be made. Instead, a simple management or automation system can be used to coordinate deployment activities.

For example, an online gaming company can use a separate management system to trigger a deployment process across clusters around the country. And each control plane retrieves and deploys the specified package independently.

- **Primary-secondary**

A primary-secondary topology can be realized with the fan out topology with a global routing policy that routes the traffic to the primary cluster by default and falls back to the secondary cluster when the primary cluster is down. A federated control plane is unnecessary in such a setting.

In burst computing scenarios, compute resources from the secondary cluster may join the control plane hosted by the primary cluster. When the primary cluster needs to burst compute into the second cluster, it uses node selectors to allocate additional instances to the secondary cluster. The secondary cluster can optionally cache scheduling decisions to achieve a certain level of autonomy, as you've seen in KueEdge.

- **Hierarchical**

This is the most complex topology with layers of control planes linked up in a tree structure. When an application is deployed, it may get cascaded down multiple levels before the actual instances are scheduled. Clusters can be assigned to different partitions or different business domains of the system. And node selectors and filters are used to make sure all components go to the designated destinations.

For example, a smart stadium may have multiple clusters, each for a specific purpose such as security control, ticketing, vending, lighting, broadcasting, analysis and many more. In a hierarchical topology, applications are pushed down from the root control plane, and application statuses are aggregated back to the central control plane for a single-pane-of-glass management view.

- **Multi-site span**

In this topology, a cluster spans multiple physical sites but remains a single logical cluster. This topology is quite demanding – it requires stable network between the clusters because the control plane assumes all participating nodes are local. The topology allows some advanced workload placement and dynamic call delegation. For example, the control plane can shift the number of instances running on different sites based on measured loads to avoid network congestion.

- Hub-and-spoke

Hub-and-spoke is a two-layered hierarchal topology, in which a leading cluster controls multiple scattered subsidiary clusters. In this case, the leading cluster wants to assert stricter control over other clusters. For example, an application rollout can be triggered only from the leading cluster.

It depends on the topology of your system, you need to pick different federation strategies – do you want a unified control plane, or synchronize control plane states through data synchronization? Do you need each cluster to operate autonomously when disconnected from others? Do the clusters assume different roles, or are they simply shared compute resources? Different answers to these questions led to different federation design.

5.6.2 Key challenges

Managing a federated environment is complex regardless of if the environment is on cloud or on edge. Furthermore, because edge devices are often scattered or loosely managed, managing federated environments has some additional challenges, specifically: discovery and bootstrapping, routing and traffic management, workload and data mobility and security.

- Discovery and bootstrapping

Unlike the cloud environment in which compute resources are carefully managed, in an edge environment, a compute resource may join or leave the compute plan at any time. When a new device joins, it needs to be properly identified, authenticated and provisioned. Then, it needs to acquire (or be given) required software stack with necessary patches. And then, it needs to establish secured communication channel with other devices or cloud. Because of the heterogeneous nature of edge, this process is quite complicated. We'll discuss the bootstrapping process in more detail in Chapter 7.

- Routing and traffic management

Edge networking environment is dynamic and often constrictive. For example, many edge networks don't allow inbound traffic. Exposing edge-hosted services to the public is a challenge by itself. When workloads roam on the edge, consistent, reliable routing is also challenging. Fortunately, SDN offers great help in addressing these challenges.

- Workload and data mobility

Edge workloads are often mobilized for various reasons – to optimize performance, to reduce cost or to simply follow where the user is. For example, as a user wearing a smart watch jog on a trail, she may be handed to different cell towers to provide continuous service. Such frequent movement requires a reliable and efficient way to migrate workloads across the compute plane. Moving user data along compute is more complex due to factors such as privacy concerns and extra latency caused by data copy.

- Security

Security is too big a topic to be covered by a few sentences here. There are many vectors that need to be considered – identity, data security, authentication and authorization, communication, secret management, intruder detection and many others. Even securing a Kubernetes cluster itself is a sizable task. We'll discuss security in more detail in Section 5.7.

At the time of writing, there isn't an established Kubernetes federation solution, other than the KubeFed V2 (https://github.com/kubernetes-sigs/kubefed) project. KubeFed defines a single set of APIs to manage multiple Kubernetes clusters. The key feature of KubeFed is to propagate *templates*, *placements* and *overrides* across cluster boundaries. These entities form the foundation for *status*, *policy* and *scheduling*, on top of which higher-level APIs can be designed to realize scenarios like fan-out deployments, geo-failover and geo-distribution.

5.7 SECURING KUBERNETES CLUSTERS ON EDGE

Security remains one of the top challenges in edge solutions. Although Kubernetes comes with reasonable defaults and various supporting features, it doesn't provide protection along these vectors out of box. For example, the Center of Internet Security recommends using a different certificate authority (CA) for *etcd* from the one used for Kubernetes. Because *etcd* plays such a central role on a Kubernetes cluster, using a separate CA reduces the risk of an attacker gaining access to the ETC store. This section discusses a few selected security topics. This is by no means a complete coverage, but a presentation of some important security vectors you need to consider.

5.7.1 Container isolation

A key threat model of containers is attacks from inside the container, such as by vouge code loaded into the container. Since the beginning, containers have been using Linux features to provide isolation around application processes – *namespaces* to create isolated instances of global resources (such as networking and PID), *cgroups* to limit resource (such as CPU and memory) usages and *chroot* to create a virtual root folder or the process. However, the vouge code can escape from these mechanisms by exploiting bugs through system surface such as *syscalls* and/*proc* files.

One solution to address this challenge is to reduce the attack surface of Linux kernel calls. For instance, you can use fine-grained, per-process access control using SELinux, or use Seccomp BPF (SECure COMPuting with filters) to filter system calls based on explicit policies.

Instead of using the security primitives yourself, you can leverage container runtimes with built-in sandboxing capabilities such as gVisor and Nabla to achieve stronger isolation without you having to do much. These solutions use an emulated Linux kernel in user space and use technologies like *seccomp* to secure delegated calls to the host Linux kernel when necessary. From the application's perspective, it still accesses system calls as before. And the protection is applied by the sandbox environment itself. At the time of writing, Google is also working on a *sandboxed pod* concept that creates sandboxes around pods instead of containers.

And finally, the extreme solution is to use a dedicated kernel for a container, as shown in Figure 5.3(b). This is a heavier solution but provides stronger isolations between containers because compromising one kernel doesn't affect other containers/kernels.

Finally, a container-based OS uses container runtime process as the PID1 process and runs a single container on a physical host. As there are no additional layers to penetrate, this also gives a strong isolation of containers.

5.7.2 Data protection

At the beginning of the section, we mentioned that you need to configure security on the cluster data stored in *etcd* using a separate CA. In later Kubernetes versions (1.7 and higher), you can also enable data encryption on *etcd* to protect cluster data at rest.

For application data, you typically need three levels of protections: at rest, during transition and during usage. Encryption is an effective way to protect data at rest. You can also leverage specialized hardware like Trusted Platform Module (TPM) to securely save your secrets like encryption keys and certificates. To protect data during transition, you need to use secured communication channels such as communication over TLS. Protecting data during usage is a bit complicated, as data usually needs to be decrypted and processed in plain text. Fortunately, there are technologies and algorithms to help. For example, you can use a hardware- or software-based secured enclave to provide an encapsulated processing environment. And you can use different multi-party computing algorithms to allow participants to collectively calculate outputs without revealing inputs. Finally, it's also possible to train and run inference on machine learning models with encrypted data.

Secret management is another aspect of data protection. Due to the extra sensitivity of the data, you want to get the best protection around this data. You can leverage Kubernetes default secret store (1.7 or higher starts to support encryption at rest), various cloud-based secret vault, or Hardware Secure Module (HSM) devices to save your secrets.

5.7.3 Access control

You need to implement proper access control to both your applications as well as the Kubernetes control plane. You should always enable role-based access control (RBAC) on your cluster and assign only the minimum required permissions to different components.

You should also enable network access controls, such as restricting pod-to-pod traffic through network policies or establishing explicit ingress or egress points. You can also use service meshes such as *Istio* or *Consul* to reinforce network policies without modifying your application code.

5.7.4 Security monitoring

Gaining awareness is the first step to mitigate security risks. There are many libraries and tools for you to implement security monitoring systems at different scales with different levels of sophistication, including Prometheus, Jaeger, Grafana and many others. The key thing to keep in mind, however, is that putting a security monitoring system is never the end goal. A monitoring system is just to give you data. You'll need to establish procedures to respond to the data before your system can benefit from the monitoring infrastructure.

5.8 SUMMARY

Kubernetes has become a pervasive infrastructural platform for both cloud and edge. We examined various aspects of running Kubernetes cluster on edge in this chapter. First, we introduced various extension points of a Kubernetes cluster such as CRI, OCI, device plugin and Cluster API. Then, we surveyed several lightweight Kubernetes clusters that are designed to be deployable to edge with lower demands on resources. Then, we went through each of the possible extension points and introduced several custom CRI and OCI implementations. Finally, we discussed cluster federation and some security concerns managing Kubernetes clusters on edge.

So far, we've discussed various edge compute patterns that focus on either bring cloud to edge or to connect edge to cloud, and how Kubernetes serves as a universal infrastructural platform for both cloud and edge computing. In the next chapter, we'll discuss applications that are designed for "pure" edge scenarios without cloud counterparts.

Chapter 6

Edge native design

The longest record of a human holding their breath underwater is a bit over 24 minutes. With some professional gears such as oxygen tanks, some world-class divers can stay underwater for over 30 hours. With a conventional submarine, mariners can stay underwater for tens of hours. And with the most advanced nuclear submarines, sailors can stay underwater for decades – at least in theory. As the most advanced species, we've created tools that enable us to survive under the water for essentially as long as we wish. However, we've chosen the evolution path to be land animals. We must rely on our tools to adapt for the underwater environment. On the other hand, tens of thousands of marine species have been living freely underwater for thousands of years without any of our fancy gears. They are native residents of the ocean; we are just visitors.

If you view edge as an ocean, you'll find some applications trying hard just to remain afloat, and some other applications carrying heavy life-supporting gears to be operational. At the same time, you'll see schools of other applications roaming and thriving freely in the ocean. We call these applications *edge native applications*. And they are our primary interest in this chapter – we'll discuss how to identify edge native applications, and how to create one.

6.1 EDGE NATIVE APPLICATIONS

To test if something will float on water, we can simply drop it in the water and see if it remains afloat or not. To test if an application is an edge native application, we can put the application into an edge environment and observe how it works or crashes. When we do this for enough times, we can start to identify some common characteristics that make an application edge native. This section offers some of our observations.

6.1.1 Autonomous bootstrapping

A device may have stayed on a shelf for a long time before it's turned on. When it does get turned on eventually, it needs to first establish the context around it. This is called a bootstrapping, which is unnecessary in a controlled compute environment such as cloud. Obviously, a smooth and reliable bootstrapping process not only greatly improves user experience, but also reduces the overall management cost. Hence, we summarize the following characteristic of edge native application:

> *1: An edge native application supports an autonomous bootstrapping process with minimum human intervention.*

Bootstrapping is a complex process, especially when you try to design a generic, fully automated process. We'll discuss zero-touch provisioning in more detail in Chapter 8. Meanwhile, we'll present a seemingly simple problem to illustrate levels of complexity you may get into even for solving such a simple problem.

You may have noticed that many connected devices require you to set up system date, time and time zone early in the bootstrapping process. This is necessary because otherwise the handshake with backend service will fail because of the clock skew. There are vastly different ways to help a user set this up:

- By providing a user interface (UI) for users to enter the settings.
- By connecting to a public time server to automatically sync time settings (through Network Time Protocol, for instance).
- By getting time settings from your router, given your router has already been properly configured.
- By installing an onboard battery to keep preconfigured settings available.
- By getting time from a global positioning system such as GPS or BeiDou.

And when you choose your solution, you must consider various constraints such as cost, form factor, resource consumption and connectivity.

Given all these, your bootstrap process is very likely to be highly customized for your specific product. This works fine for a single product line. However, this generates a big problem for a generic platform trying to provide a unified or fully automated experience. We'll present some ideas on how to design such a system in Chapter 8.

6.1.2 Adaptive to environmental changes

As we introduced in Chapter 1, comparing to the cloud environment, edge computing environment is heterogeneous, constraint and dynamic. This requires an edge application to be ready to adapt to environmental changes, such as a network becoming disconnected, a power supply running low, environment temperature rising above the safety threshold, peered devices becoming unavailable or an assigned certificate is about to expire. While applications running on cloud, many of these environmental changes are addressed by human operators or automated scripts in a timely fashion because computing hosts are closely monitored. The edge environment is far more challenging. In many cases, a known problem won't be addressed for a long time – just recall how long it took the city to fix a broken streetlamp outside your house. For me, that was about 3 months. Because of delayed corrections, edge applications often need to adjust their own behaviors to extend their life spans before the reinforcement arrives. A good example of this is a cell phone switching to low power mode when its battery is running low. We observe the second characteristic of an edge native application as follows:

2: An edge native application adapts to environmental changes and uses multiple alternatives to deliver required capability.

The second part of the above statement requires more explanation: What we mean is that an edge application is well prepared for possible environmental changes. It does this by providing multiple ways to deliver required capability. And it can switch among these methods based on environmental changes. For example, it's critical for an emergency radio to remain operational during extended blackouts. Therefore, some emergency radios are designed with multiple power sources – rechargeable battery, solar panel and a hand crank. Another example is a data collecting sensor using a local cache to store limited amount of data while connection to the backend service is broken.

Obviously, building multiple implementations of the same capability incurs additional cost. You'll need to decide on how many implementations you bring into your device or application based on your service-level agreement (SLA) goals, cost targets as well as business justifications (such as if the multiple implementations differentiate your products from those from your competitors).

Not all environmental changes are for the worse, though. A native edge application also knows how to take advantage of resources around it when possible. To achieve this, the application needs two fundamental capabilities: to discover new resources and to make use of the discovered resources. Dynamically discovery of new resources and capabilities is a very interesting topic that deserves much more than a few sentences here. We'll discuss the topic in great detail in Chapters 7 and 8 when we introduce the capability-oriented architecture (COA).

6.1.3 Edge high availability

For a cloud service, the easiest way to achieve high availability (HA) is to deploy multiple instances of the service behind a load balancer. These instances are often scattered across multiple *fault domains* so that failures are contained in smaller scopes instead of propagating to the entire system.

Before we continue, we need to give a proper introduction of a fault domain. A fault domain (or a failure domain) is a logical or physical segment of a compute plane that will be negatively impacted when a critical device or service experiences problems. For example, a desktop PC can be a fault domain in your house. When any of its components – monitor, hard drive or keyboard – fails, the entire system becomes unusable. However, a failing computer doesn't have impact on your other appliances such as dishwashers and toasters. In a cloud datacenter, a service rack is often a fault domain. This is because each of the rack has its own power, cooling and networking components. Failure in any of these critical components renders the enter rack unusable, but it doesn't affect other racks in the same datacenter.

Redundancy within the same fault domain is often less useful than you may have thought. If the fault domain is controlled by critical devices or services without redundancy, having multiple instances of your applications doesn't help when these critical pieces fail. So, when you design your edge HA solution, you need to make sure your critical infrastructural components are backed up by redundancy (so that you have multiple fault domains) before you worry about redundancy of your applications.

For example, when you deploy a Kubernetes cluster on the edge, you need to make sure the control plane is deployed on to multiple nodes, preferably with independent power supplies, to reduce the chance of a complete control plane failure. Then, you deploy your applications with multiple instances running on multiple physical nodes so that a single failing node won't bring down the whole application. This brings us to the third characteristics of edge native applications:

   3: *An edge native application designs for high availability across multiple fault domains.*

Unfortunately, redundancy across physical nodes isn't always feasible. As we explained in Chapter 1, edge computing is *computing in context*. An application instance can't failover to another node if that will lead it falling out of context. For example, an application controlling USB-attached devices can't simply failover to another node because the devices won't be accessible from any but the current node.

In such cases, you may need to consider building up redundant infrastructures within the same context, as shown in Figure 6.1(a). The diagram shows that the infrastructure is split into two independent fault domains. As long as one of the domains is available, a downstream application (not shown) can continue to operate by getting data stream from either domain. Figure 6.1(b) shows an improved design with a messaging layer inserted between the sensors and the processors. There are a few benefits of this design: (1) The processors can be deployed outside of the physical context; (2) The size of fault domains is reduced, constraining the scope of negative impacts caused by errors; (3) Processor A and processor B serve as backups for each other because they can pick up each other's messages from the message bus. This also allows the two processors to split the workload while both being healthy.

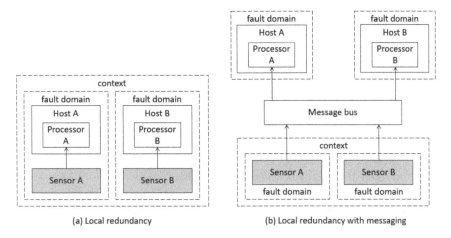

(a) Local redundancy   (b) Local redundancy with messaging

Figure 6.1 Local redundancy architectures.

Last but not least, auto-restart of your applications upon power recycle or reboot is often necessary to keep availability within a single fault domain. You can achieve this by leveraging operation system features such as *systemd* for Linux and auto-restart services for Windows.

As mentioned before, keeping applications available is not only a problem when updating or upgrading the application itself, but also when you upgrade core systems such as Operating systems or runtimes. In order to guarantee zero-downtime system updates/upgrades, we also need to introduce *update domains* (scope of updates). You need at least two update domains spread across two *fault domains* (scope of failures) to guarantee a no downtime system update or upgrade. A system runtime update/upgrade will then be applied to only one update at a time.

6.1.4 End-to-end security

An application running in cloud can often reside in a *secured sub-system* that is provided by the hosting environment. For example, a web service can be protected by an application programming interface (API) management solution that provides necessary authentication. And the security context is made available to the application through attribute claims carried by security tokens or augmented requested headers. The application can then use the information to make appropriate authorization decisions. At the same time, application data is kept safe in carefully monitored databases with authentication, role-based access control (RBAC), restrictive firewall rules, encryption at rest, periodical backups and full audit records. Furthermore, cloud provides additional productions such as network group policies, anomaly detection and protection against DDoS (distributed denial-of-service) attacks.

Protecting the edge computing environment is much tougher. Among everything else, a very practical threat comes from the fact that an adversary has easier access to compute resources on the edge – we've seen in movies how robbers simply smash the surveillance system to wipe their traces on their way out. In addition, the more devices that are connected to networks the more attack surface hackers have, it's as simple as that. A real-life example is to tap optic fiber by bending the fiber at specific angles and reading leaked data using some cheap hacking devices or the example where a casino was hacked through an unprotected thermostat in a fish tank which allowed hackers to steal 10 gigabytes of high-roller user data. And those are just a tiny set of examples out of many other creative ways to exploit hardware access. Furthermore, in many cases devices are not the final targets of hackers. Their goal is to use compromised devices to attack your backend system, through when they can cause serious damages.

To implement strong security, an edge device or application should always assume breach and uses a defense-in-depth strategy to build up layers of protections around computing resources and digital assets. Figure 6.2 illustrates a possible defense-in-depth strategy that uses a combination of software, hardware, physical protection as well as reinforced business process to protect the edge computing environment.

We summarize the following characteristic of a secured edge application:

   4: An edge native application assumes a breach and applies a defense-in-depth strategy to protect compute resources and digital assets.

When it comes to security, it's always easier said than done. Very few of us are true security experts. One practical suggestion we can offer is to leverage as much of professional services or tools as you can – such as hiring an independent firm to perform periodical security reviews and audits.

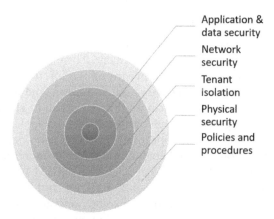

Application & data security

Network security

Tenant isolation

Physical security

Policies and procedures

Figure 6.2 A defense-in-depth strategy.

The following list is a collection of principles that you should understand and consider in your solutions:

- Threat modeling along your solution stack is always necessary. Implement security features comes with its associated cost. By understanding your scenario context, you can make informed choices on what security mechanisms you put in place. For example, in a smart home scenario, as the thermometer data is streamed directly into a collocated hub server, transmitting data in plain text is probably fine because the probability of intrusion is low. On the other hand, in a smart building scenario, you will need to consider the case if a hacker or a terrorist tries to trigger automatic fire sprinkler by tampering with the sensor data. We'll discuss how to model your edge solutions in a consistent stack and present a threat model in the next section.
- For devices processing sensitive data, it's necessary to establish the *root-of-trust*. A root-of-trust is the point that is presumed unbreakable. And the whole trust chain is originated from the root-of-trust. For instance, a hardware root-of-trust can securely boot up the system and ensure only certified code (signed or attested) is loaded for execution. It also establishes root keys for encryption and often offers a trusted execution environment to execute critical instructions in isolation.
- Because devices work in real-life contexts, the privacy implications brought by such attachments can't be ignored. A good principle to follow is *minimum exposure*. This principle requires that a party must have justified business reasons to access any piece of data. For example, it's justifiable for a shipping application to get a user's address. However, it's often unjustified for the application to get the user's date of birth. A hacker often collects scattered data from multiple sources to form a complete profile of his targeted victim. So even if you consider having the extra data passed around doesn't present an imminent security threat to you, you should still limit the amount of transmitted data for the sake of your customers' privacy.
- Observability and manageability. The importance of observability is amplified in edge environment because of the open, dynamic compute plane. Collecting telemetries and operation logs from your devices will help you to paint a clear picture of the overall health of your edge system. Your devices and applications should always be patchable, either locally or remote updates. Although remote manageability isn't always feasible, you need to at least provide capabilities to cut off data stream of a vouge device (from your data collecting service, for example).
- Leverage machine learning (ML) solutions for advanced protections. With recent development of machine learning, more and more machine learning models are available for advanced detection such as anomaly detection and DDoS detection. And with advanced in transferred learning, chances are you can partially leverage existing mature

models and tailor them for your specific needs. It can be anticipated that the cost and complexity of implementing a ML-based solution will go down, and using sophisticated ML is not a far-fetched goal even for medium-to-small projects.

6.1.5 Manageability at scale

Managing a single Internet of Things (IoT) device is easy. Managing tens of IoT devices is troublesome. Managing hundreds of IoT devices or more is very difficult. Managing large number of devices requires not only technical readiness but also other important aspects such as business process, training, scheduling and more. In this section, we'll focus on how to evaluate the manageability of an edge system. And in Section 6.2, we'll cover how to keep manageability in mind when you design your edge native applications.

A key metric for measuring manageability is the mean time to repair (MTTR), which reflects the average time for a problem to be detected, isolated, analyzed and resolved. Problem detection relies on a reliable health monitoring and alert system. The alert system can be measured by precision (formula 6.1) that measures the accuracy of the positive detections and recall (formula 6.2) that measures sensitivity of the detection system.

$$Precision = \frac{True\ Positive}{True\ Positive + False\ Positive} \quad (6.1)$$

$$Recall = \frac{True\ Positive}{True\ Positive + False\ Negative} \quad (6.2)$$

A key to improving manageability at scale is automation. Automation reduces human errors and ensures consistency in update processes. For systems with long MTTR due to complexity of repairs (such as fixing elevators or other complex machineries), you should consider establishing a maintenance system that predicts possible failures and schedules proactive actions to amend the system before an actual failure occurs.

At this point it is worth discussing another interesting aspect of manageability, automation and dynamic environments when considering a desired state and the actual state.

Manageability usually entails automating the provisioning of a platform stack. Infrastructure as code solutions like Azure Resource Manager or Hashicorp's Terraform allows the user to define the desired state of an environment in a declarative or imperative way. This allows for continuous, automated and consistent creation of resources in an idempotent way. One can apply the same approach to edge environments where the desired state of a single device or an entire environment is captured by such a model.

As we have seen in previous chapters, digital twins (DT) are a way to retrieve the actual state of a device or an environment and that a DT needs to be modeled and connected to the physical environment. It is not hard to see that the desired and actual state can become of sync quite easily due the dynamic nature of some edge environments. This can either be triggered by adding/removing new devices (not to be confused with adding another instance of a machine/service, we are really talking about topology, configuration and software changes) or by changing the device/environment based on user context, where the desired state could be downloaded at startup.

Most available twin services enable one to reason over static rules applied to state changes but do not support more dynamic approaches needed such as when states of multiple twins are no longer in a desired state. This is further complicated, not just by time-specific rules for a dynamic state, but by a desired state to be a dynamic set of twin states (aggregated, calculated, etc., with time as a variable too) that is reasoned over to arrive at this computed desired state.

At the time of writing there was no real solution for this and one needs to address the problem by "manually" keeping states in sync, but we wanted to mention it as it is something that needs to be accounted for when working with a desired state and DT solutions.

We summarize the manageability characteristics as:

> 5: *An edge native application uses automated monitoring and a patching system to ensure continuous operation.*

Finally, version management is an important aspect of manageability. We'll discuss effective version management in Section 6.2.

6.2 A MODEL FOR DESIGNING EDGE NATIVE APPLICATIONS

Edge native applications are challenging for various reasons: First, an edge application is most likely a distributed application. And distributed applications are among the most complex computer applications. Second, managing an edge application often involves managing the full hardware and software stack. This differs from cloud-based applications that can leverage cloud infrastructure (IaaS) or platform as a service (PaaS) as managed hosts. Third, edge applications often have demanding performance and security requirements with strict resource constraints and assumed deadlines. Finally, an edge native application often involves multiple original equipment manufacturer (OEM) vendors, protocols and services. And the hardware and software stack often change over time.

We propose a one-stack-multiple-perspective (OSMP) model that helps you to maintain clarity throughout the life cycle of your edge native application. We believe the OSMP creates a stable framework for planning, designing and implementation. OSMP is a guidance system that helps you to navigate through the complexity of an edge native application. And it's designed to be used in conjunction with other software engineering processes and design patterns, which are subject to change based on project requirements and preference of specific teams.

6.2.1 The OSMP model

The OSMP model is inspired by the 7-layer network model, which has been proven to provide a common abstraction over a huge number of hardware and software systems. We use a similar approach and define a simple linear stack that captures the entire hardware and software portfolio of an edge native application. Our initial proposal of the model is illustrated in Figure 6.3. The stack is a complete stack that covers the entire hardware-software spectrum, from the hardware layer all the way to orchestration at the top. So, the first functionality of the OSMP model is to remind you not to miss any mandatory layers when you decide on the scope of your project. Then, based on the same model, you can create different perspectives that help you to focus on different aspects of system design and operation. Possible perspectives include provisioning, day-2 operation, messaging pipeline, threat model, performance optimization and others. Because these perspectives are applied to the same stack, the OSMP model helps you to

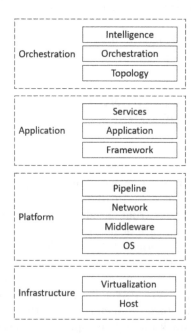

Figure 6.3 The OSMP model.

align technologies chosen in different perspectives to ensure that working on one perspective has minimum negative impacts on other perspectives.

The OSMP model favors virtualization. Especially, it assumes network capabilities are delivered by software-defined networks (SDNs). Therefore, *network* is put under the platform layer. This choice aligns with 5G network design principle, which emphasizes SDN functions over generic hardware. For example, Cloud-RAN (CRAN or C-RAN) is a cloud radio access network that aims to use cloud to satisfy increasing demand on 5G. It separates the traditional base station into a radio part and a baseband part. The radio part remains in the base station in the form of a remote radio head (RRH) unit and the baseband part runs on cloud in a centralized and virtualized baseband unit (BBU) pool running on generic cloud platform hardware.

The OSMP model doesn't explicitly define interfaces among different layers, because it's impossible to settle for a limited set of interfaces for all kinds of edge applications. However, when using OSMP model, it's important to explicitly define interfaces among different layers, as these interfaces define *management domains* that divide the complex stack into isolated, manageable pieces that are better aligned with typical team member expertise. Hence, we recommend keeping clear consensus and maintaining as much as stability on these interface choices as possible throughout the life span of your project.

The OSMP model provides stability as well as a converging point across different teams and work streams. Based on the same OSMP model, you can define multiple perspectives that focus on different aspects of the project. For example, you can have a *provision* perspective and you can use it to check if you've get necessary solutions lined up to provision required hardware and software pieces. Figure 6.4 clearly lists out what needs to be

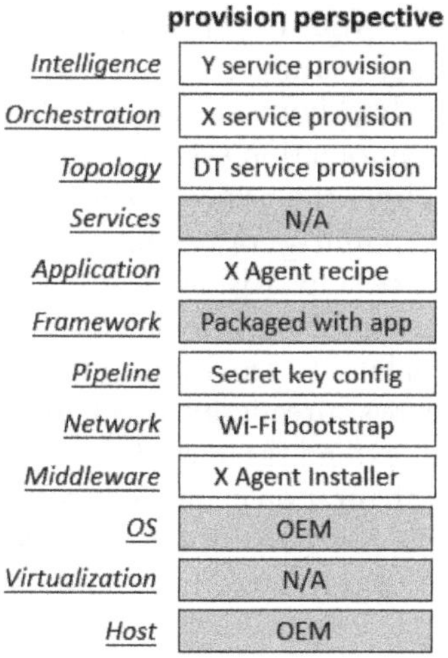

provision perspective

Intelligence	Y service provision
Orchestration	X service provision
Topology	DT service provision
Services	N/A
Application	X Agent recipe
Framework	Packaged with app
Pipeline	Secret key config
Network	Wi-Fi bootstrap
Middleware	X Agent Installer
OS	OEM
Virtualization	N/A
Host	OEM

Figure 6.4 A sample provision perspective.

done at each layer so that the team has clarity on what's supposed to happen across all layers.

The above sample shows a simple IoT scenario in which an edge-based sensor collects data to a (cloud) managed X service. The X service ships with a client-side agent (the X agent in Figure 6.4) that can deploy user code through a user-defined recipe. The system topology is designed using another managed DT service. And system insight service is offered by yet another managed Y service. Because the system uses managed services as well as a system with built-in capability of distributing application bits, the provisioning process isn't complicated. However, Figure 6.4 reveals that there are still individual pieces that need to be taken care of. Having such clarity will help you to avoid loopholes and surprises in your project life cycle.

You may also want to extend OSMP for your specific project needs. For example, in the proposed model, data pipeline is presented as a single layer. However, if your solution requires a scalable data pipeline, you may want to expand the pipeline into individual pieces such as *ingress*, *processing*, *storage*, *visualization* and *insight*.

6.2.2 Perspectives

You can use the OSMP model to track multiple perspectives at the same time with some simple tools such as a spreadsheet. However, you can do more with the spreadsheet than just tracking individual perspectives. The following are some sample usages when you examine the spreadsheet in different ways:

- Ensure alignment. When you check across the rows, you may find different teams have made conflicting or redundant choices. These are opportunities for you to step in to get teams aligned with their technical choices. For example, two different teams may have chosen different messaging backbone, different orchestrator, different frameworks or different automation systems. Getting them to converge on fewer options simplifies the project complexity.
- Seek opportunities to unify. For example, you may have observed that Figure 6.4 shows service provisioning in the first three rows. In such cases, you may consider using a templating system such as Terraform or Azure ARM templates to provision all the services as a single action. Similarly, given there's already an agent installed on edge, you may consider to reuse or extend the agent for other client-side delivery jobs such as pushing configuration changes or software updates.
- Checking across the rows again. If you notice that most of the perspective cells are *N/A* or offered by a vendor, you may want to work with your vendors to eliminate the rest of the cells so that you can cleanly cut off good chunks of the project and hand them to your vendors and partners.

The concept of perspectives allows the OSMP model to be horizontally extended to cover different aspects of the project. How the perspectives are chosen is at your own discretion. Regardless, the spreadsheet gives you a clear map of all the moving parts for you to make informed decisions.

6.2.3 Views

Once you have your stack and perspectives defined in the same spreadsheet, you can create multiple views on top of the same spreadsheet. Visualization is always a great tool in project management, especially when you manage a complex edge computing application. For example, you can create a dependency graph view to help you to monitor cross dependencies, to identify critical paths and to optimize work schedules. Or you can create a simple green-light/red-light view to give you a bird's-eye vision over the progress of the entire project.

If you were using Microsoft Excel, you can create multiple tabs. You enter the OSMP model and your perspectives into the first tab, and you can create multiple views (tabs) using cell references to the first tab. This allows you to create as many views as you like while keeping a single set of master records. Of course, you can still use project management tools of your choice to track your project and use OSMP as a map to help you to navigate.

The last point we want to make on the OSMP model is that it provides a possible convergence point across your edge computing projects. Given the heterogeneous nature of edge computing, we don't think there will be a single established technical stack. And the diverse technical stacks have been problematic for tool creators to create universal tools that will work across different edge computing projects. The OSMP model creates a possibility to converge on a consistent *workflow*, which can serve as the stem of a set of universal tools that works across a great spectrum of edge computing projects.

6.3 EDGE NATIVE DESIGN

Before we continue, let's recap the key characteristics of an edge native application:

- 1: An edge native application supports an autonomous bootstrapping process with minimum human intervention.
- 2: An edge native application adapts to environmental changes and uses multiple alternatives to deliver required capability.
- 3: An edge native application designs for high availability across multiple fault domains and possibly update domains.
- 4: An edge native application assumes a breach and applies a defense-in-depth strategy to protect compute resources and digital assets.
- 5: An edge native application uses an automated monitoring and patching system to ensure continuous operation.

These five characteristics cover both architectural aspects and operational aspects of an edge native application. Yet we still consider these as merely starting point. As you dig deeper, you'll discover many more details to be considered. For example, #2 characteristic states that an edge native application should be adaptive. Writing an adaptive application itself is a very challenging subject. This is why we covered OSMP first – we hope the model can help you to stay on track as you go deep into each aspect.

6.3.1 Bootstrapping

Formally, the bootstrap process includes a commission phase and a provision phase. Commission is the earliest phase of a managed entity. Some actions, such as identification, certification, time configuration, etc., must be performed before the entity can be provisioned. Then, the entity needs to be enrolled with a backend system through a series of actions including discovery, identifications, establish trust and software deployment.

With proliferation of end-to-end IoT systems, the provisioning process is increasingly segmented in edge computing. Essentially, almost every system requires you to install some sort of agent on devices and to establish device identity with the backend system. And these systems are often opinionated in communication protocols, manifest formats and device capacities. So, the backend system you choose has profound impact on various parts along the OSMP stack. Furthermore, many such systems assume a flat topology, in which devices are directly connected to the cloud. If you are designing a hierarchical topology, you'll also need to think about how to fit the hierarchical topology into a flat management structure – such as by using tags and virtual device groups.

There is not much innovation design you can apply at this stage, because you need to adapt to the backend system you need to work with. The only exception is that when you need to connect to multiple backends (such as a management backend and a separate messaging backend), you can think of ways to orchestrate the individual bootstrapping processes into a unified flow. The authors of the book have been discussing with a few partners on practicality of implementing a generic "bootstrapper of bootstrappers" to handle such scenarios. Reach out to the authors if you were interested in the effort.

6.3.2 Adaptive design

Adaptive design (not to be confused with adaptive UI design) has three components: data collection, decision-making and a call proxy. In such as design, telemetry data of interest is collected and fed into a decision-maker. And the decision maker dynamically reconfigures a call proxy to use appropriate candidate service provider based on the decision logic. When a call request arrives, the proxy invokes the configured service candidate and returns the result, as shown in Figure 6.5.

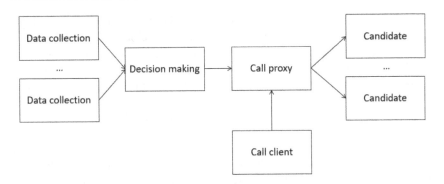

Figure 6.5 Adaptive architecture.

It's worth noting that in the previous design, decision-making is not on the call path. This works well in common cases that adaptive actions are infrequent events (think of how many times you need to switch between online/offline mode), because there's no per-call overhead to pay. However, if your scenario requires per-call decision-making, you probably want to merge the decision component into the call proxy itself so that the proxy can make dynamic, fine-grained routing decisions.

Another possible choice is to set up the entire reactive mechanism at a lower level such as the network service mesh level: you can feed collected data to service mesh to drive dynamic routing rules. The benefit of doing this is that you can manage your decision policies at runtime without modifying the code. And you can apply consistent policies across multiple applications that share the same set of backend services.

If your application offers some services to be consumed by a downstream caller, you may consider to offer an endpoint for your client to query available services, and level of services you are offering at the moment based on the current context. For example, your camera service may offer a colored stream by default, and a night vision stream when lighting condition changes. By providing a query API, you allow downstream services to better adapt to your service changes.

We'll discuss how to design adaptive applications using COA in the next chapter.

6.3.3 Edge high availability

Reliability, availability and serviceability, which are often collectively referred as RAS, are three pillars of a reliable and available system. Essentially, availability ensures the system is up and running, reliability ensures the system does what it's supposed to do and serviceability ensures a broken application can be fixed with minimum downtime.

As discussed earlier, redundancy has been a key technique to provide high availability for cloud-based solution. It works for cloud because cloud has a huge resource pool that can be leveraged to acquire compute resources as needed to replace the broken ones. Although redundancy is still a key technique for edge in many cases, it's not universally applicable due to various constraints. In such cases, improving reliability and serviceability becomes very important to improve availability of individual compute resources without redundancy.

Improving reliability and serviceability is a huge engineering topic that can't be sufficiently covered here. Instead, we summarize the following principles that are most relevant to edge environment as a quick guide when you make architectural decisions in your edge application design. This is not an exhaustive list for sure. It serves mostly as a reminder of some useful design tools that you should keep in your toolbox.

- Build alternatives for your critical functions. This is an effective way to build up redundancy within a single compute unit. For your critical functions, you want to have alternative implementations to avoid single point of failures. For example, although you may prefer a remote management interface, you should consider a local management interface as well, in case you lose access to the remote interface.
- Explicitly version everything. As your system evolves, it's always helpful to explicitly version your service contracts and your data contracts. This is especially necessary because your edge deployments often span across multiple application versions. Having explicit versioning will allow you to recreate and isolate problems more easily.
- Enable self-service when possible. Enabling self-service is a great tool to reduce servicing workloads. If you provide prescribed methods for your users to diagnose and fix common problems themselves (such as resetting passwords), you can shave a huge chunk of maintenance work from your shoulders. In advanced scenarios, you can also consider allowing the system to be put into a diagnose mode that an end user can use to analyze the system state and collect debugging information for further analysis.
- Enable to restore to factory default settings when possible. This is your last line of defense before going through an expensive hardware servicing process. However, resetting to factory default settings may lead to customer data loss in some cases. If your user data is periodically backed to cloud, you can try to restore to the last known working state with the latest snapshot.

As introduced in Chapter 5, lots of effort exist to enable Kubernetes to run on a constrained environment. You should consider those solutions when appropriate. On the other hand, we hope the previous list can give you some ideas on how to add on the simple redundancy approach.

6.3.4 End-to-end security practices with the OSMP model

In edge computing environment, end-to-end security must cover everything from hardware to software and the communication between devices and cloud. When you use OSMP model, you can examine security layer by layer along the stack, and you can also examine security from different perspectives. The model gives you a map to do certain level of threat modeling yourself. And even if you hired specialists to model the threats for you, you can also use the OSMP to ask correct questions to ensure every corner is covered.

You can go a step further by creating an OSMP security view that encompasses common threat perspectives, including confidentiality, integrity, authentication, availability violation and privacy. Then, you can go through the stack and analyze threats from common attack venues such as network-based attacks, software attacks, hardware attacks and insider attacks. History has taught us that insider attacks are often among the most devastating attacks. A RBAC policy and Just-in-time access mechanism are proven technologies to alleviate such threats.

6.3.5 Manageability

What makes managing edge solutions more complicated than managing cloud-based environment is the need to manage the whole life cycle of the whole stack, from the commission and provision all the way to decommission. Again, you can use the OSMP model as a map so that you can track various management activities throughout the project life cycle.

To ensure manageability, you often need to implement in-band (IB) management interface as well as out-of-band (OOB) management interface. IB interface is visible to the software and firmware running on a system. For example, a simple IB management thread sends heartbeats signals to a management system. And missed signals will trigger the management system to perform corrective actions, such as restarting the application. The OOB interface is often implemented independently from the system to be monitored. This allows the system to be diagnosed and repaired even if it's powered off or corrupted.

Manageability at scale often boils down to two fundamental problems – secret management and package distribution. Once these two problems are addressed, we can deploy any trustworthy software packages and use the secrets to establish private connections back to the management system. When you design your manageability subsystem, you should first ensure you have reliable, scalable secret and package distribution channels because they are the foundation of everything else in the subsystem.

Edge computing is quite complex. It's impossible for us to provide a comprehensive guide with half of a chapter. We barely scratched the surface here. But we hope the presented ideas, especially the OSMP model, can inspire some thoughts on designing, developing and operating edge native applications.

6.4 SUMMARY

In this chapter, we discussed key characteristics of an edge native application. Then, we proposed an OSMP model that can be used as a map when you navigate through the complex landscape of edge computing. And finally, we covered a list of principles that you can follow when you attempt to design your own edge native applications.

We've analyzed different design patterns, frameworks and tools for developing edge computing solutions in the past six chapters. In the upcoming two chapters, we'll introduce our creation – COA, which is a combination of observed pattern and proposed framework to tackle complex edge computing systems.

Capability-oriented architecture

Chapter 7

Introduction to capability-oriented architecture

One of the professors in my college was very popular among the students. He had a little cardboard pie chart on his door with a needle. On the pie chart were different places he might be – dining hall, classroom, library, etc. Whenever he left his office, he turned the needle to point at where he was going so that students could find him pretty much at any time.

In analogy to computer science, the pie chart served as a service discovery mechanism – it told the students where to find the professor to establish a connection with him.

When the headmaster wanted to meet with him, it was a different story. The headmaster would send some faculty members or students to find the professor and bring him to the headmaster's office.

In a nutshell, that's what capability-oriented architecture (COA) i about – to fulfill your intention (meeting with the professor) by bringing the required resources to you, instead of you going around searching for the required resources. And the pivotal component of COA is a capability proxy.

7.1 CAPABILITY PROXY

Instead of hunting down the professor himself, the headmaster uses a student or a faculty member as a *capability proxy*. He tells the proxy that he wants to meet with the professor, wherever the professor might be at the moment. While the proxy scurries around trying to acquire the professor, the headmaster can focus on more important stuff, such as sipping coffee. Figure 7.1 illustrates how such an interaction happens using the capability proxy. The proxy acquires the professor *somehow* and offers the headmaster an *IMeeting* interface, through which the headmaster can meet with the professor.

In the context of edge computing, the capability proxy encapsulates all details of acquiring the requested interface and offers the interface to the consumer as a local call. Behind the scenes, the proxy might be connecting to a remote service, or downloading and launching a Docker container instance, or delegating calls to a peer on the same network. The consumer of the interface remains oblivious – from its perspective it simply calls

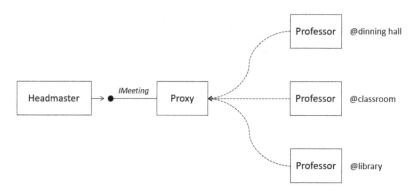

Figure 7.1 A headmaster using a capability proxy.

a local interface that is always available. Edge computing happens in a very dynamic compute plane. The capability proxy brings simplicity and stability to the chaos, hence making designing and implementing the capability for consumers much easier. And because the interface may be delivered by different mechanisms, we call the exposed interface a *capability* instead of a *service*. This is not just a name change but something significant. Capability represents an abstract functionality. It isn't concerned with how the functionality is delivered. This abstraction is the root of many innovations in COA, which we'll cover in this and the next chapter.

Let's return to the capability proxy for now. The capability proxy can do much more than just acquiring and exposing a capability to the consumer as a local interface. It enables many powerful patterns that are particularly relevant to edge computing scenarios.

7.1.1 Location transparency

The capability proxy can be configured with multiple *offers* of the same capability. And it can dynamically choose and switch among offers based on policies, constraints and operation conditions. For example, an object detection capability can be backed by a cloud-based offer with more sophisticated but more expensive AI model and a local offer with a lightweight AI model that offers less detection confidence. The proxy calls into the cloud-based AI model by default. However, when the network is disconnected, it switches to the local model to ensure business continuity. It can be even made adaptive to network bandwidth changes – when the bandwidth is high, it sends high-resolution pictures to the server; when the bandwidth is low, it switches to low-resolution images.

All these are hidden from the consumer. It has no idea the proxy is working hard to provide an adaptive, highly available capability to it. All it does is to call the object detection interface. The proxy also separates developer concerns from operational concerns. The operation team can decide what offers are behind a capability, and design policies and constraints as they see fit for the particular deployment. The developers that consume the capability don't need to care about any of those.

The capability proxy works with multiple offers to deliver a capability. This arrangement enables some quite powerful hybrid patterns, as introduced below.

7.1.2 Hybrid patterns

Because offers of the same capability can come from anywhere – cloud, local network, same host or even in-process calls, the capability proxy can be configured to enable some very useful patterns:

> **Circuit breaking with failing over to local.** Circuit breaking is a common pattern in a service-oriented architecture (SOA). When a service has an extended outage, it doesn't make sense to keep trying to call the service. Instead, a circuit breaker "trips" after a certain amount of attempts to avoid futile attempts. The capability proxy can be configured to act as a circuit breaker. However, when the breaker "trips", it's also capable of failover to a local offer to provide continuous service to the consumer.

> **Bursting to cloud.** A capability can favor local offerings by default for lower cost and higher performance. However, when the call volume exceeds the local processing power, it can burst into cloud by launching or requesting cloud-based processing units to take over the excessive workload. And when the traffic dies off, it switches back to the local offerings to avoid unnecessary cloud hosting and invocation costs.

> **Caching.** The capability proxy can cache capability invocation results in a local cache. This means the proxy is able to serve repetitive calls directly from its own cache instead of calling the downstream service upon every request. This behavior not only lessens the burden on backend servers, but also offers faster response times when there's a cache hit. This works quite like a content delivery network (CDN). The difference is that in this case, the proxy can cache arbitrary compute results in addition to static contents. And it has greater flexibility to refresh its cache on a per-call basis instead of a fixed policy. For example, in a truck fleet management system, a proxy may choose to pass truck location queries directly to the backend when the truck is on duty and serve cached last-known location while the truck is off schedule.

> **Dynamic delegation.** Edge scenarios often involve devices with drastically different processing power. The capability proxies sitting on the same network can discover and communicate with each other. And they can delegate tasks to each other assuming prior consensus from the user. In my living room, the most powerful computer is my XBOX

One console, which is loaded with high-end GPUs. When some other smart devices need additional processing power, they can in theory delegate complex calls to the XBOX. For example, my front gate camera can run sophisticated facial recognition (instead of detection) on my XBox and then delegate a voice warning to a capable device such as my Amazon Echo Dot to verbally tell me that my wife is at the front gate.

Assemble multiple offers. If you think the previous scenario is pretty cool, you should hear the following scenario I've been telling my friends almost like a joke. Imagine you want to find your TV remote. You talk to your Echo Dot to find your remote. Echo Dot broadcasts a location query to connected devices. My Kinect picks up the request and locates the TV remote using its cameras. Then, Echo Dot generates a request to deliver the remote to you. Your robot vacuum takes over, drives to the TV remote. A little motor in your remote vibrates and it hops onto the vacuum. The vacuum talks to a camera (such as the Kinect) to locate you and delivers the TV remote to you. I always get a good laugh when I tell the scenario. But I don't think we are that far from the science fiction. The key here is that no single vendor has designed this remote delivery capability. The devices negotiate with each other and jointly come up with a feasible solution. That's fascinating, isn't it? Although we won't be able to cover all of the details of how this works, we'll explain the main ideas of how this can work in the next chapter.

We've barely scratched the surface here. Later in this chapter, you'll see how the capability proxy integrates with other parts of COA and delivers interesting features such as selecting offers based on non-functional requirements and contributing compute power back to a distributed compute plane. But before those, let's spend some time discussing a less glorious but equally important feature for enterprise edge computing scenarios – middleware.

7.1.3 Proxy middleware

As reflected in Figure 7.1, the capability proxy is architecturally a *sidecar* of the consuming application. Because the sidecar intercepts all incoming and outgoing calls of an application, it can be used to deliver many cross-cutting features without modifying the application code, such as distributed tracing, compliance policy reinforcement, service throttling, data normalization, API version adaption, protocol translation and many more.

The capability proxy is designed to support a middleware system, through which the aforementioned features can be plugged in as *middleware components*. A middleware component can intercept inbound traffic, outbound traffic or both. Middleware configuration is once again an operation concern that remains transparent to developers. Figure 7.2 illustrates a capability proxy with pluggable middleware components.

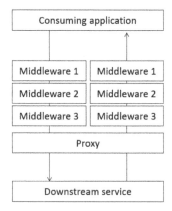

Figure 7.2 Capability proxy with middleware.

Figure 7.2 seems a bit complicated. We need to ask the question that how such a software stack is distributed and installed on edge devices. This is done by a proxy bootstrapper, which is described in the next section.

7.1.4 Proxy bootstrapper, capability set, capability profile and acquisition agent

A bootstrapper pulls down and installs the capability proxy on an edge device. Under the COA architecture, a capability is described by a *capability profile*, which defines the exposed capability application programming interface (API) as well as candidate *offers* that can be used to deliver the required capability. An offer describes details of how to acquire a consumable endpoint of the offer. This can be as simple as a Hypertext Transfer Protocol (HTTP) endpoint, or as complex as a custom *acquisition agent* that acquires, installs and configures a local instance for consumption.

An optional *profile monitor* can be deployed to monitor capability profile changes and triggers a reconciliation process that brings the capability profile to the desired state. This may involve calling the acquisition agent to acquire a new version of the capability. The acquisition agent is designed to be self-upgradable so that it will update itself if the update capability profile requires a new acquisition agent.

The capability profile also describes a *strategy* of choosing offers. For example, a strategy may prefer cost, while another strategy may prefer service-level agreement (SLA) associated with the offer. An enterprise policy may also dictate to use offers from only approved vendors. All these non-functional requirements can be factored into the strategy to influent how the capability proxy makes choices among available offers.

Multiple capability profiles are organized into a *capability set*, which describes a set of capabilities available on a device. Both capability proxy

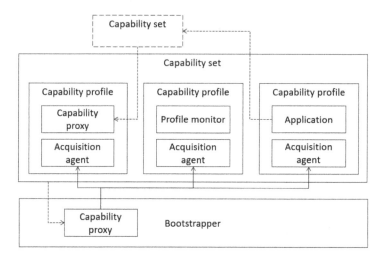

Figure 7.3 A bootstrapper on a device.

and profile monitor can be described by capability profiles in the set. And the capability proxy can be configured with additional capability sets to acquire additional capabilities. And because all the bootstrapper does is to material-ize a capability set, so the bootstrap itself can be implemented by a capability proxy. This kind of recursive definition may sound confusing, so let's try to clarify that with a diagram. Figure 7.3 illustrates how it looks on a device with a bootstrapper. The bootstrapper is configured with a capability set that contains three capabilities: a capability proxy, a profile monitor and a custom application. The bootstrapper works with corresponding acquisition agent to acquire the necessary software packages and launch them. Then, the application feeds its required capability set to the capability proxy. And the capability proxy in turn does exactly what the bootstrapper does – to bootstrap required capabilities for the application. Hence, the bootstrapper is nothing more than a capability proxy that takes in a capability set and uses acquisition agents to install and configure required software stack.

This is a very extensible design, because it defines a façade of capability acquisition without settling with any particular protocol, package format or package repository. The design also supports software updates through the acquisition agent, including updating the acquisition agent itself.

Once we solve how to bootstrap the capability proxy onto a device, we need to consider how to make the proxy highly available for continuous operation, which we'll discuss next.

7.1.5 Proxy HA

We define capability proxy high availability (HA) as a capability proxy that should have the same or a longer life span than the consuming application.

When an application runs as a loose process, the proxy is configured as a system service that is auto-launched when the device starts and is auto-restarted when it crashes. There's no device failover in this case – if a device fails, everything on the device – the application and the proxy – fail. This doesn't violate our definition of HA though, because the proxy has equal or longer life span than the consuming application.

Furthermore, the acquisition agent can be separated out as an independent service. It forms a *watch pair* with the proxy. The two services watch each other. When one fails, the other agent will attempt to restart it. So, the pair fails only when both agents fail.

When deployed on a cluster with built-in HA scheduling capability such as Kubernetes, the capability proxy can be injected into the application pod as a sidecar container. This gives the proxy the exact life span of the relying application. When the application pod restarts, or fails over to another node, the proxy comes with it.

This concludes what we have to say about the capability proxy for now. The proxy is a key concept in COA. It provides a highly available, location-agnostic sidecar that delivers required capabilities to an application. The application can be designed with abstracted capabilities while all capability delivery mechanisms are abstracted away from developers.

We've been mentioning capability as an abstract concept. However, how do we describe such a concept? And how do we achieve a balance between simplicity and expressiveness without scarifying precision? These are the questions to be discussed next.

7.2 INTENTION

Going back to our headmaster sample – the headmaster has a very simple intention, which is to meet with the professor. Although looks trivial in the real-life example, realizing such communication pattern in computer science is difficult. Our first hurdle is to express our intention. I'm not even talking about natural language processing here – just any means for us, or our applications to express a clear intention, such as the simple intention to add two integers (as defined by mathematics instead of representations in computer science).

Well, we have means to find a method that is named *add()* or *sum()*, and we can even examine the method signature to see that it indeed takes two integers and returns a new one. However, we have no guarantee that it will actually add two numbers. On the other hand, when you tell elementary students to add two numbers, they know exactly what to do. What's missing here is the ability to understand the semantic meaning of the demand, which is a difficult task for computers and machines.

One way to solve this is to use an API or interface with implied semantic meanings. All participants agree on the semantic meanings beforehand and interpret the API in a consistent manor – this is actually what's happening today. When you call an API, you are implicitly buying into the semantic meanings it represents. From COA's perspective, using traditional API as a way to describe intention is totally acceptable. On the other hand, we do wonder if we can offer a dynamically discoverable mechanism that allows a consumer to establish the semantic meaning on the fly.

7.2.1 Interpreting user intention

Fortunately, over the years the industry has been studying ways to solve the problem in limited domains. If we have a well-defined syntax, a well-known lexicon and clearly defined semantic algebra with valuation functions, we can convert a sequence to word into a precise operation in a predefined domain. Furthermore, for describing complex data types, we have created various data schemas that describe precise layouts of data packets. However, let's not go there just yet. For now, we'll "cheat" and try to make this work with some radical assumptions:

- "Add" is a well-known word that all parties understand. It represents an arithmetical add operation on two given numbers and returns a new number.
- "Integer" is a well-known data type that represents a 64-bit signed integer.

Then, the intention of adding two numbers can be expressed by a simple sentence: "Add 2 integers". Similarly, if we consider "meet", "professor" and "name" as well-known words, the intention of meeting the professor can be expressed as a simple imperative sentence – "Meet a professor whose name property is Philip". Equations 7.1– 7.9 capture the basic context-free grammar (CFG) that can be used to interpret such intentions (assuming "integer" is already defined).

$$P \rightarrow VP \tag{7.1}$$

$$VP \rightarrow V\ NP \tag{7.2}$$

$$NP \rightarrow Dsc\ N\ Pre \tag{7.3}$$

$$Dsc \rightarrow \text{'}one\text{'} \mid \text{'}two\text{'}\ \#\text{ignore in path} \tag{7.4}$$

$$N \rightarrow (integer) \mid \text{'}professor\text{'} \tag{7.5}$$

$$Pre \rightarrow \text{'}with\text{'}\ PN = PV\ \#\text{as parameters} \tag{7.6}$$

$$PN \rightarrow \text{'name'} \tag{7.7}$$

$$PV \rightarrow \text{'} \langle identity \rangle \text{'} \tag{7.8}$$

$$V \rightarrow \text{' add'} \mid \text{'meet'} \tag{7.9}$$

Once we have the established grammar, we can use it to translate word sequences to structured constructs, on top of which we can build semantic understanding of users' intentions. And we can also use more advanced ways such as natural language processing (NLP) to translate user sentences to structured intentions.

Why do we go through these troubles? Remember, our goal is to allow developer to design application using capabilities that are not bound to any specific implementations. Hence, we need a language to describe capabilities. And we want to make sure these capabilities are meaningful to the consumers. So, we define a *capability* as an activity that can *fulfill* a user's *intention*. To do this, we establish a common lexicon with a list of verbs and nouns. A verb in our vocabulary has explicit semantic meanings, or mapping from input range to output domain. There are two types of verbs: *fundamental* verbs and *complex* verbs. COA defines a finite list of fundamental verbs that carry out various state manipulations such as creation, transformation, deletion, aggregation, segmentation, identification and match. The intention described by a verb phrase with a fundamental verb is a *basic* intention. An *advanced* intention is a net of basic intentions. And it can be described by verb phrase with a complex verb. A complex verb is any verb as long as an advanced intention can be decomposed to only basic intentions. For example, "bake a pizza" is an advanced intention with a complex verb, bake. This advanced intention is decomposed into a network of basic intentions, all described with fundamental verb phrases, such as:

- **Combine** water and flour → dough
- **Combine** pepperoni and dough
- **Change** oven temperature attribute to 400°F
- **Place** dough into oven
- **Wait** for 25 minutes
- **Retrieve** dough from oven

This decomposition is called a *plan*. A plan has one or more steps, with each step fulfilling a *sub-intention* by executing the corresponding verb phrase. In other words, a simple verb phrase generates a one-step plan, while a complex verb phrase generates a multi-step plan.

How big is the basic verb vocabulary? We don't know for sure at this point as it's still to be fully defined. But we estimate it won't be a big list – probably under 100. The complex verb, on the other hand, is an open vocabulary. The "meaning" of a verb is reflected by a plan. And participating parties are assumed to have consensus on the plan through a shared repository or other means of consensus.

Similarly, the noun vocabulary is comprised of an even shorter list of primitives, on top of which complex nouns can be constructed using different composition strategies such as a property bag, a tree structure or an ontology.

In the real world, an intention is often associated with a context. For instance, the intent to meet with the professor has an implied time box – the headmaster probably wants to meet the professor soon instead of a year later. Computers don't natively understand these contexts. Hence, we need a way to explicitly tell computers various contexts and constraints associated with an intention.

7.2.2 Intention annotations

An intention can be annotated with a series of key-value pairs. These key-value pairs are compared against the similar key-value pairs associated with an offer when the capability proxy decides if the offer is a valid candidate. The keys and the values belong to the same shared lexicon that is mentioned in the previous section.

These annotations can be used to capture various non-functional requirements, such as (but not limited to) vendor qualification, security requirements, latency constraints, cost limit, locality preferences, minimum SLA guarantee and other contextual information.

Annotation mapping can be simple or complex. In a simple case, a strict one-to-one mapping is checked and only an offer with perfect match is considered. More complex mapping rules can be designed, such as range mapping, fuzzy mapping and combined mapping with logic operators. When there are many annotations, the mapping problems can be further extended into a linear constraint problem or even a machine learning model that captures interleaved connections among these values.

Once we allow clients to express their intentions, we can enable a new kind of service discovery – *semantic service discovery.*

7.3 SEMANTIC SERVICE DISCOVERY

Semantic service discovery is used to identify qualified offers by semantic intentions instead of calling syntaxes. In the dynamic edge computing environment, the ability to perform semantic service discovery enables powerful scenarios. For example, a smart home scenario often involves different devices from multiple vendors. And a user rarely plans ahead what devices to install in the house over time. So, new devices can be introduced into the little ecosystem, and older devices may be removed at any time. Although leading appliance vendors try to create ecosystems around their own brands, interoperability across brands is usually problematic. Furthermore, because an appliance is often designed to deliver a unique set of capabilities in isolation, complex and dynamic use cases spanning multiple devices are often considered add-ons or nice-to-haves. With semantic service discovery and intention decomposition, more complex scenarios can be enabled, even when none of the device vendors have anticipated them.

Imagine you are working in the kitchen and someone rings the doorbell. You ask your voice assistant such as Microsoft Cortana, Amazon Alexa, Google Assistant or Apple Siri, "Who's at the door?" What the device will do is to decompose the intention to a series of basic intentions, and "ask around" to see who can fulfill them: who can take a picture, who can identify a face in a picture and who can map a face to a contact. And this "asking round" process is done through semantic service discovery.

7.3.1 Semantic service discovery process

To make itself semantically discoverable, a service needs to either register itself with a *service registry*, or to respond to a broadcasted query. For example, a router with semantic discovery capability can serve as a well-known service registry to all attached devices. It offers a predicable discovery endpoint to all devices attached to it, and it syncs its settings with a central registry that is managed by the network carrier.

A service registration contains the capabilities the service delivers, as well as any associated annotations. The semantic service discovery process matches the user intention with service capability descriptions and annotations and takes matching records as candidate offers. The matching process can be precise or fuzzy (by synonyms, for example), hence an identified offer can be associated confidence value that indicates how confident the system thinks the offer can fulfill the intention. As the capability proxy formulates the plan to fulfill the intention, it should take confidence values into consideration when making choices among offers.

Once a candidate offer is identified, the discovery process runs an optional step to confirm with the service if it is willing to make the offer. This allows a service provider to proactively refuse offering a service in certain cases, such as it's too busy to take on new requests, it thinks the ask price is too low, or it can't verify the origin of the ask.

A confirmed offer resolves to one of the following things: a callable capability endpoint, or a traditional service discovery endpoint, a web service description, or a *recipe* to acquire a service instance.

7.3.2 Capability endpoint

A capability endpoint is a service endpoint providing a COA capability API. The API defines callable routes, which are formed by concatenating tokens in a COA grammar definition (such as what's defined by Equations 7.1 through 7.9) by route separators. For example, a capability of adding two integers can be invoked through path:

```
/add/integer
```

And the route for meeting with the professor is:

```
/meet/professor/with?name=philip
```

Some tokens are marked to be ignored in path building (such as Equation 7.4), and some tokens are marked to be built as query parameters instead of path parts (such as Equation 7.6). You can observe both cases in the previous examples.

The payload to the capability endpoint is a single JSON-encoded data structure expressed by nouns defined by the COA lexicon. And the return object is also a single JSON-encoded data structure made up by the same set of nouns. Annotations are passed as request headers (such as HTTP headers) or metadata (such as gRPC metadata). For example, the payload to invoke the "add" capability is simply:

```
[100, 200]
```

We realize that such a uniformed syntax may become hard to use in some cases. Hence, we allow ways to connect to "classic" service discovery and consumption mechanisms, as introduced in the next section.

7.3.3 Classic service discovery and description

The semantic service discovery process may return a classic service discovery endpoint, through which the client can use traditional service discovery mechanism to discover a service endpoint. It may also return a service description, such as a WSDL (Web Service Description Language) document, to instruct the client on how to consume the service.

Just like COA doesn't mandate a specific API interface to be exposed by the capability proxy, it doesn't mandate a specific format or proxy how a capability is discovered or consumed. Specific implementations may make different choice. Converging on a standard format such as what's described in Section 7.3.2 is obviously preferable to improve interoperability. However, we don't anticipate everyone converging on the format. So, COA shall remain open and flexible. This allows the community room to jointly build up consensus and gradually converge on unified grammar, protocol, serialization format and other things.

7.3.4 Acquisition recipe

Semantic service discovery may return a *recipe* to acquire an ability. For instance, a recipe can contain instructions to pull down a Docker image from a Docker registry and to launch a local Docker container to provide required service. Another example of a recipe could be a shell script that uses Linux package manager to download and install required binaries and to launch a new system service.

A recipe defines a desired state as well as how to reach the desired state. This doesn't necessarily mean the recipe has to do everything, though. As a matter of fact, because there are many state seeking systems out there, a recipe can simply leverage one (or several) of such systems to reach the desired state.

We won't formally discuss decommissioning a capability here. However, our general idea is to keep a list of acquired capabilities and to evict the oldest (or least frequently used) capabilities when necessary – such as when the system is running low on disk space. The decommission instructions can be packaged in the recipe as well – in this case, the recipe contains instructions to seek a *null sate*, the state of missing certain capabilities.

Figure 7.4 recaps the semantic service process with different routes a capability proxy may take to fulfill a requested capability.

Before we conclude this segment, we'd like to briefly discuss different dynamics between a client and the semantic discovery service. Since our goal is to allow a client to intelligently expand itself as needed, we need to empower the client to make smart choices of which services to call. The next section presents a client-initiated auction process that illustrates how client can assert controls on what services to consume.

7.3.5 Client-initiated auctions

Everyone knows that when you about to make a serious purchase, such as a TV, you'd better shop around before you make a decision. Even if you are not

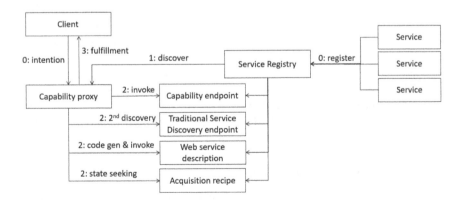

Figure 7.4 Semantic service discovery process.

an expert of TVs and don't know what to look for, you can educate yourself by observing how features and performance differ as you compare among different brands. You'll learn terms such as HD, UHD, HDR, OLED, etc.; you'll experience how pictures look with different sizes, refresh rates, resolutions and color profiles; and you'll also appreciate different design styles that may or may not match your living room. Based on this, you can build a list of "must-haves", "desirable", "optional" features. And eventually, you choose the one that can satisfy all of your "must-haves" and preferably some of the "desirables" while trying to stay within your budget.

COA allows a client to shop around before settling for a specific offer. For example, when a client tries to discover for an AI service, it can demand offers provide at least 98% recall and cost no more than 5 cents per inference. In other words, the semantic service discovery process can be viewed as a client-initiated auction. When multiple offers are available, the client can use the process to pick the best offer to get the most value with the lowest possible cost.

Because a client may acquire a local copy of a capability, it can participate in auctions initiated by other clients and offer the capability back to other clients. Over time, this discover-acquisition-offer cycle builds a dynamic, distributed compute plane and all participating clients can continue seeking optimized capability offers. We'll discuss this dynamic compute plane in more detail in the next section.

7.4 COA COMPUTE PLANE

A COA compute plane is comprised of compute units that deliver different capabilities. Under COA, there are no clear boundaries between servers, clients, clusters or machines. All compute resources on the compute plane can provide services to others. And they can continue to seek better and closer offers that can fulfill their intentions faster. This is a continuous optimization process. And as the edge compute environment changes, the process makes the system adaptive to the dynamic environment. This is quite extraordinary. It's a very complex task to manage a dynamic compute environment. Yet, by clients seeking to maximize value for themselves, a global optimization may emerge without a centralized, coordinated optimizer. This reminds me of a story I've heard – an architect was asked to design an optimal path across a campus between the dormitory and the lecture hall. He planted grass between the buildings. And as students made their way between the buildings, their footsteps collectively marked an optimal way through the grass. Then, all the architect needed to do was to pave the path. For interested readers, please research "complex adaptive systems" for more details on how such emergences work.

7.4.1 Skynet-style computing

I'm very fond of the idea of Skynet in the Terminators series. And I firmly believe that a ubiquitous, global connected, decentralized compute plane will eventually happen. And COA might provide another nudge toward that direction, as it allows devices to negotiate with each other to carry out complex actions without central coordination. The downside of such a system is that it will be pretty hard to shut it down completely. Then, as the system seeks optimization based on its own logic, there's a practical danger that some of the decisions may be made not necessarily in the human race's best interests, with or without self-awareness.

To mitigate the risk, we can design an agent whose sole task is to traverse the compute plane to keep things in order. And yes, I mean an agent like agent Smith from the Matrix. And of course, there shall be Neo whose mission is to find opportunities to update system architecture to the next revision.

I think I should stop here on the topic as I want to avoid leaving you an impression that all my reasonings are based on science fiction movies. However, I still believe the ability for an agent to dynamically extend itself and to automatically collaborate with nearby agents is a key capability enabled by COA for Skynet-style computing.

We will turn the craziness dial down a bit for a while and talk about a more practical problem – workload scheduling.

Nowadays, a common approach of scheduling workloads is to use a centralized workload scheduler. Compute nodes report their availabilities to the central scheduler. On the other hand, users submit workloads to the central scheduler as well. Then, based on compute node availability and workloads, the schedule calculates an optimum scheduling solution. The schedule algorithms can be complex, but a common greedy algorithm, a cost reduction process or an auction-based process can often generate fair results.

This kind of centralized scheduling isn't feasible on a large, dynamic compute plane that is comprised of millions or even billions of devices that may come and go at any time. Here, we propose a decentralized scheduling scheme in which each service is represented by an autonomous agent whose sole task is to seek the best place to host the service. Instead of a mathematical model, we'd like to offer some intuition here using an analogy of quicksilver droplets flowing on an uneven terrain. Image the compute plane is represented by an uneven terrain with peaks and valleys. The peaks are where the resources are strained, and the valleys are where idle resources are available. When a workload is to be scheduled on the compute plane, it's like a droplet of quicksilver that is randomly dropped onto the terrain. As droplets have a tendency to flow downward to the lowest level, the valleys are gradually filled up and we approach a global optimum. As the terrain changes, the droplets flow around to keep the average resource consumption even.

If we zoom into a droplet, we'll find the droplet is made up of one or multiple particles connected by rubber strings. When the particles are scattered, these strings try to pull them back together. This mechanism allows multiple interconnected services to be kept in proximity as much as possible.

When a droplet is used by a client, the spot of the terrain generates a magnetic field that holds the droplet in place. The hold is stronger as the

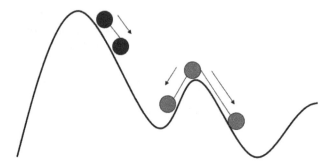

Figure 7.5 Quicksilver workload scheduling.

usage is heavy. Otherwise, the hold is weaker and can be broken by other droplets. This mechanism reduces movement of actively used services while allowing less used services to flow around held ones.

Figure 7.5 provides a simple illustration of two droplets flowing on a compute plane. As you can see, they have a tendency to move toward less occupied compute resources while trying to keep their inner parts together.

We hope the previous text and illustration provide some intuition on how such "quicksilver scheduling" works on a large, dynamic compute plane. At the time of this writing, we haven't finished formal mathematical modeling or large-scale simulations. For interested users, please reach out to the authors; we are certainly seeking help, especially if sophisticated fluid dynamics and machine learning can be applied to the problem.

7.4.2 Crowdsourced cloud

A compute node on a COA compute plane can contribute capabilities it has acquired back to the compute plane. This forms the foundation of the edge native cloud that we introduced in early chapters. The edge native cloud is a real challenge to the traditional, centralized clouds, especially in consumer-targeted services and edge computing scenarios.

Let's take web browsing as a simple example. Although the Internet is massive, an individual usually access only a few of favorite sites daily. If the in-house router is smart enough to prefetch and cache content the user is likely to access, it can provide a very responsive web experience to the user.

What if the neighbor shares the same interests?

Imagine this – as the neighbor's router looks around for required capability (that delivers contents from specific sites), it finds the router and asks if the router is willing to offer the capability. The route says, "of course, but with a fee". The two routers reach an agreement and the transaction is recorded in a distributed ledger such as a Blockchain-like ledger. Then, the transactions are settled monthly so that the capability provider gets paid for its service.

There must be burning questions now: How to protect the privacy of the owner of the first router? Should the transaction be viewed as reselling of the content? All valid questions. And we are afraid we don't have perfect answers. Regardless, this model of gaining financial rewards by contributing compute capability back to the compute plane is quite interesting, especially to telecom carriers. When more subscribers contribute their devices back to the compute plane, the carrier can host large-scale services on edge devices that are right next to the consumers. This kind of scale and reach give the carriers significant competitive advantages over the classic cloud platforms. Furthermore, the added cost of maintaining such an open compute plane is minimum, nothing much than what the carriers already do. This is a tremendous advantage comparing to billions of dollars spent by the cloud platforms to maintain centralized datacenters. Consumers go through the telecom backbone to connect to the cloud backbone. Cloud offers right off the telecom backbone and edge devices will have faster response time than making round trips to the cloud backbone, and this is decided by physics.

A practical challenge of using crowdsourced cloud for enterprise is security. It could be a hard sale to persuade enterprises to load their sensitive workloads on to arbitrary compute nodes running at unknown locations. However, there's hope. As introduced in earlier chapters, many Internet of Things devices are building up a hardware-based trusted execution environment that allows a piece of code to be attested and executed within a protected environment without interruptions from anyone – including rogue administrators. There are also systems that leverage secured multi-party computation that allow multiple parties to jointly perform computations over their inputs while keeping those inputs private. Or, when there are abundant compute resources, we can repeat the same computation on multiple (and possibly geographically separated) devices and accept the answer only a consensus is reached.

The crowdsourced cloud maybe a dream or a nightmare of hackers. The dynamic nature of the compute plane may make it harder for a hacker to target at specific targets. We are not security experts but we can imagine that a denial-of-service attack is harder to pull off because computations are scattered and sometimes happen in a peer-to-peer fashion. On the other hand, a hacker can use the dynamic compute plane to host vogue code that is harder to track down and eliminate.

What we are trying to say here is that we realize the potential problems, and we don't have proven answers other than some general ideas.

Crowdsourced computing is now a new idea. Projects like Seti@Home and Folding@Home aim to use scattered personal computers to perform distributed computations. Crowdsourced cloud differs in two ways: First, it aims to run open workloads instead of a specific type of workload. Second, it offers a financial model for compute resource contributors to gain monetary rewards or benefits. We can even imagine a new type of digital currency that is based on Proof of Contribution instead wasteful Proof of Work, which is commonly used in digital currencies such as Bitcoin.

7.4.3 Better computing for everyone

Because the COA proxy can automatically handle per-transition billing (as a middle plugged into the pipeline, for example), it can be extended to enable powerful scenarios such as spending/accumulating charity credits associated with compute tasks and enable consumer-as-provider billing through digital currency.

We also envision design principles proposed by COA can help the world to achieve more economic compute for everyone. Imagine when your COA proxy tries to discover a candidate, it factors in factors such as greenhouse gas emission associated with the compute resources and picks the greener option. This will encourage service providers to be more considerate to environment protection. Furthermore, the discovery process can assert latency and resource consumption constraints so that the most efficient algorithms get their fair market shares. The industry's collective efforts will make compute more sustainable for everyone.

We also hope COA can help to make computing fairer across the globe, by making sophisticated, efficient computing available to everyone on a shared infrastructure that is not under control of the few global enterprises. Of course, realizing such a vision takes much more than an architecture paradigm. Regardless, we wish COA will play a positive role in reaching that goal.

7.5 CONTEXT

We'll close this chapter with a brief discussion of *context*. We'll consider three types of context: user context, activity context and device context. Contexts are very important to provide a continuous and natural user experience. We'll talk about some basics here and we'll dive deeper in the next chapter.

7.5.1 User context

As a human user navigates through their environment, they may interact with a number of devices – their phone, laptop, car, TV and others. If these devices could recognize that they are being used by the same user, they can offer a continuous, personalized experience to the user. Many consumer smart devices nowadays require authentication, through which the devices can establish user contexts. And these devices gradually gain a deep understanding of user preferences by observing user behaviors. For example, most streaming services use recommendation engines to promote contents that users are likely to consume. This kind of intelligent behavior has certainly improved user experiences in many cases. On the other hand, they've also raised privacy concerns. User context is also subject to abuse. For example, an unethical ticketing site may artificially raise the ticket price for a user because his payment history suggests he can afford tickets at higher prices. Similarly, a service provider may silently mask low-priced options to lead the user to buy more expensive services as the record shows they have chosen more highly priced options before.

With technological advances, it's possible to quickly establish user context on public or shared devices as well. As you approach a store, its camera can recognize you by facial identification and notify the clerks about your preferences. In a less obvious case, a smart scale can easily identify individuals in a household by weight differences.

When properly used, user context can be the powerful glue and catalyst that enable powerful scenarios with seamless transitions among devices. In the next chapter, we'll discuss a couple of scenarios that leverage user context while respecting user privacy.

7.5.2 Activity context

One of the major failures in modern user interface (UI) design is browser favorite link management. Till today, favorites in modern browsers are still organized as a simple list. Although you can organize favorites in folders, we see the feature rarely properly used.

If we take *activity context* into consideration, we can find opportunities to improve. Imagine a typical case – you do a Google search on a topic, and you find a couple of links that you want to save as favorites. It's easy for a browser to automatically put these links into a folder using the search keyword as the folder name. And we can go further – by aggregating multiple consecutive search topics, the browser can create a bigger cluster of related links, and by aggregating by time, the browser can help you to recall things that you remember seeing a while ago. Figure 7.6 shows a mockup of a different browser favorite link manager UI that leverages activity context to automatically cluster related links by topics and time. And this is just a small sample of creating new and improved user experiences by taking activity context into consideration.

Nowadays, modern workers are frequently interrupted by incoming information, such as emails, social network messages and meeting requests.

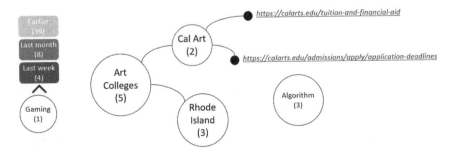

Figure 7.6 Mockup of a new browser favorite link manager.

And users often need to switch among multiple applications to accomplish tasks at hand. A context-aware COA system allows workers to deviate from the main work to handle unexpected interruptions and quickly regain focus afterward. It also allows works to stay in context while being able to access all relevant capabilities around the world.

An activity context may involve multiple user contexts. Such activity contexts can be called *collaborative activity contexts*. Scenarios such as multi-player gaming, collaborative editing and online meeting are all instances of collaborative activity context at work.

We'll discuss how activity context works in the next chapter.

7.5.3 Device context

Device context aims to enable a device to become aware of the surrounding environment. An obvious example is a device that is adaptive to network conditions, as we discussed earlier in this chapter. A device can be made adaptive to other environmental factors, such as battery level, lighting condition, wind, temperature, acceleration and others.

It's interesting to think about how a device would react to low battery conditions. If it discovers a method to acquire more power (such as by moving toward a wireless charger) through COA, will it decide to ignore user tasks and make getting power a priority? In other words, will the device have an instinct to survive (by sustain power supply) that overrides human-assigned priorities? Humans build survival instincts through evolution; a machine can be given such instincts by encoding a nonnegotiable rule to maintain power supply. Instincts generate motivations. Motivations generate decisions. Decisions generate actions. Hence, device context enables devices to act as living, intelligent entities that are driven not by preloaded machine learning models but by reactions to environment changes. We tentatively call this kind of behavior *artificial motivation*. We'll not expand on this topic further in this book, but we will continue to study device contexts in the next chapter.

7.6 SUMMARY

COA is an architectural paradigm for ubiquitous computing. Due to the dynamic and heterogeneous nature of ubiquitous computing, an application often needs to be designed to be adaptive to environment changes. COA abstracts the complexity away and makes capabilities available as local services, regardless of how the capabilities are hosted and delivered. COA also formalizes concepts such as intentions, semantic service discovery and contexts to provide reusable design patterns, as we'll discuss more in the next chapter.

Chapter 8

COA applications

Have you ever installed a smart phone application so that you can just do one thing from a specific service provider? Looking at my own phones, I have an Uber app to get a ride, a Caviar app to get delivered food, a Lugg app to get help to ship a furniture, a Delta app to check in to Delta flights, along with a few others.

What if you want to get a ride from a different company, or check in to a different airline? You need to get yet more apps. What if there's another company that you don't know of but offers rides at a lower rate?

What if, we don't have any applications on our phones?

8.1 A PHONE WITHOUT APPLICATIONS

Imagine you are planning a dinner night with your friend. You search the Internet to find a good restaurant. Then, you use a booking site, or the restaurant's site (or app) to make a reservation. And then, you search for a florist who can deliver a bouquet to the restaurant. Once everything is booked, you'll share the booking with your friend through an email app or a messaging app. Eventually, when the time comes, you will use another phone app to get a ride to pick up your friend and then to the restaurant. If you were doing all these online or through phone apps, you are likely to use at least 3–4 apps and you have to make sure things are matching up across the apps. If you ever want to change something, you have to manually make sure everything is shifted properly, including informing your friend about schedule changes – believe me, you may forget. It happens.

If we infuse the browser with capability-oriented architecture (COA), we can provide direct actions based on activity context. Figure 8.1 shows a mockup user interface (UI) of a COA-enabled browser. As you click on a restaurant link, commonly used capabilities are made available in context as contextual menu. You can book a table, reserve a car, buy a bouquet and send the invite to your friend without leaving the page.

If you want to do something that is not listed as menu items, you can use a dedicated gesture to bring up a command box in which you can type

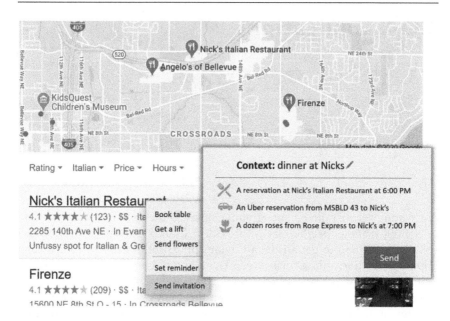

Figure 8.1 COA-infused browser.

in your intention, or to open the microphone so that you can dictate your command.

With the help of activity context, these capabilities are orchestrated so that everything happens at the right time at the right place. And for each service you use, you are not bound to a specific service provider. You'll be able to automatically discover and choose the best service options during the time at given location. There are no multitasking or context switches. You can perform all these actions right on the spot.

In the above scenario, the browser is the entry point of the whole workflow. However, with COA, it doesn't matter where the entry point is, because you always have access to all possible capabilities regardless where you are. The browser can be launched as an application. Or, in the world without applications, the browser is launched as a capability to fulfill a "browse Internet" intention. A COA-enabled mobile phone can provide a system-level gesture (such as holding on a special button) for user to trigger a new activity, such as "edit doc", "browse Internet" or "watch video". These activities can also be pinned on phone screen as shortcuts so that you can launch them without voice commands. This also gives a familiar experience to users who are used to use applications. Please note COA-enabled cell phones can support existing mobile applications as they are. And these applications are augmented by all discoverable capabilities if they were made COA-aware as well.

Figure 8.2 illustrates how COA provides a unified user experience across application boundaries so that a user doesn't need to switch between multiple activity contexts to accomplish tasks that span application boundaries.

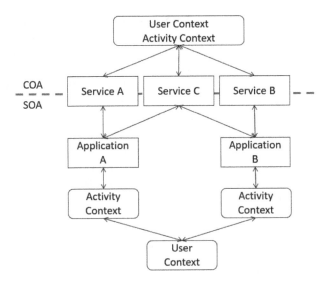

Figure 8.2 Activity context spanning multiple applications.

Devices without applications are not science fiction. If you think about it, when you interact with voice-controlled devices such as Echo Dot, you are not interacting with any specific applications. Instead, you can access any supported capabilities at any time without explicitly launching applications. And the device can establish user contexts by recognizing the person's voice. What's still lacking as of today is the ability to establish and maintain more comprehensive activity contexts that span multiple operations.

Applications are how capabilities are packaged and delivered. They don't necessarily represent the best or the most natural way of how these capabilities are consumed. We hope what we've presented here will inspire more thinking on how to design user experiences around user activities instead of applications.

8.2 INTELLIGENT APPLICATIONS

Although we've discussed devices without applications, we are not suggesting all applications should be eliminated. Instead, we think applications can provide curated experiences for complex activities such as coding, data analysis and media editing. What we propose in this case is a new genre of applications that have their foundation use cases in mind, but are able to dynamically enhance themselves, sometimes to the extend beyond what the designers have originally intended.

An application extending itself is certainly not a new idea. Many applications support the concept of *plugins*, which are additional code packages you can acquire to enrich application features. And we've observed software suites such as Microsoft Office 365 and Adobe Creative Cloud providing integration points among individual offerings to accomplish advanced operations. These plugins are designed for specific applications. And acquiring the plugins is often an explicit user action.

An intelligent application can dynamically extend itself to leverage all relevant capabilities available in the COA ecosystem. This allows even the simplistic applications such as Notepad to extend itself to provide sophisticated functionalities. For example, you can select rows of numbers in your text file and invoke an "average" capability to get the average value of these numbers inserted at your cursor location. Or you can select a paragraph in the text document and invoke a "translate" capability to translate the text into a different language.

Now, let's consider a more complex scenario – you've selected a few rows of numbers and invoked a "charting" capability to plot a bar chart of the data. The result of the capability in this case can't be inserted back to the original editing context. Instead, the activity context needs to be extended to include a new application.

Figure 8.3 shows a fictional UI framework that allows the newly involved application to push the original application aside. And the user can return to the original application by a designated gesture (such as swiping left). The figure also shows a previously invoked "print" capability fading to the right as you return to the charting application.

The key difference between this UI framework and the classic multitasking UI is that the activity context is maintained as a glue to link the involved application together. A user can swipe left and right within the same activity context. This helps the user to remain in focus because she doesn't need to manually hunt down the correct windows. And the data transfer among applications is transparent – the user doesn't need to copy and paste data through system clipboard. Instead, data transfer is carried out by the capability proxy (who may leverage the system clipboard behind the scenes).

With this kind of design, it's hard to tell where an application starts and where another application ends. And this is precisely the point – we want to

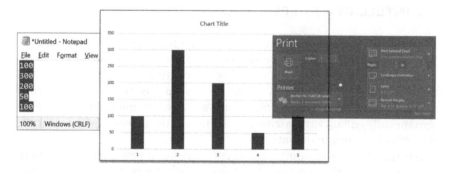

Figure 8.3 A UI framework that supports activity contexts across applications.

drive user activity to completion using all available capabilities, regardless which applications are providing the required capabilities. Recall the activity shortcuts in the previous section – they can be used to explicitly launch new activity contexts. Additional gestures, such as swiping up and down, can be designated to switch among activity contexts.

COA can help to improve the existing application ecosystem by separation of labor. With recent AI developments, intelligent features such as object detection, optical character reader (OCR), natural language processing and facial recognition are becoming expected utility features instead of key differentiators. However, training and hosting an effective machine learning model is not a simple matter. Especially, to provide continuous operation when a network is disconnected; an additional mechanism is needed to distribute and update machine learning models on clients. Furthermore, to tailor a machine learning model for a particular use case, you need ability to continuously collect new data and retrain the model. All these additional concerns go beyond what service-oriented architecture (SOA) describes and can be encapsulated by the COA paradigm.

COA also helps backend service providers to get into the mobile ecosystem. Creating a mobile application, especially a native mobile application requires knowledge of UI design, mobile user interaction convention and mobile platform integration. And to publish an application, the creator needs to go through the application store onboarding process as well as application verification and publication process. This is a skill set that many service providers don't process (nor do they care). COA allows these backend services to be dynamically discovered and consumed by all mobile applications (comparing with explicit, static integration under SOA) without the providers taking on the burden to maintain mobile applications themselves. If the mobile platform provides an integrated capability proxy, all applications can use the proxy to consume various capabilities. On the other hand, the proxy can provide built-in billing support so that the capability providers can be rewarded. Since COA doesn't tie applications to specific capability providers, the capabilities are motivated to continuously improve their services because they have constant opportunities to grab business from others while facing the threat of losing their own market shares.

8.3 ZERO-TOUCH DEVICE PROVISIONING

Device deployments often start with the process of bootstrapping devices with the backend management system. Although seems trivial, this bootstrapping process is complicated when carried in a secured, automated fashion at scale. Before we discuss how COA can help to create a generic device provisioning system, we first review how device provisioning happens.

8.3.1 Device provisioning process

Before a device can join an edge computing solution, it needs to establish its identity with the backend system and establish a secured communication channel to exchange data and configuration with the backend system:

1. Device distribution and power up

Devices are preloaded with a **bootstrapper.** The bootstrapper can come in different forms, such as firmware, boot loader, kernel, initialization process, automated script or auto-launched agent. The bootstrapper's mission is to move the device from an *uninitialized state* to a working state. And once the device is configured, the bootstrapper is no longer used. Some devices support *factory reset*, which wipes device data and reverts the device into the *uninitialized state*, which reactivates the bootstrapper.

2. Acquiring an authentication key for provisioning

For secured provisioning, the bootstrapper needs an authentication key for the provisioning process. This key is used to establish a secured connection with the **provisioning system,** which is often different from the backend to which the device eventually connects. A device can be preloaded with a certificate that is trusted by an optional well-known **discovery endpoint,** through which the device can locate the provisioning service. Some other devices require user intervention to establish the device identity with the provisioning service. For example, upon power up, some devices put themselves as access points (AP). And a user uses a mobile app or a browser to manually enter necessary information – such as a service set identifier (SSID) and password to a Wi-Fi network – for the device to establish a connection to the provisioning service. Please note, the connection in this case is not secured. The device can use the preloaded certificate to establish a secured connection. Or the network router can be configured with a preloaded certificate and open access to the discovery endpoint or provisioning service. This allows devices on the local network to connect to the remote endpoints without per-device certificates. Lastly, some devices can directly establish identity with the backend service. In such cases, separate provisioning keys are not required. For example, a subscriber identity module (SIM) card can securely store the international mobile subscriber identity number and related key to establish device identity with the mobile network carrier.

3. Enrolling with the provisioning service

Once the device connects to the provisioning service, it can establish its identity with the provisioning service. This can be a manual process or an automated process. In a manual process, a human user enters the device identity that is distributed through an out-of-band method, such as a serial number or bar code printed on the device. In an automated process, the device identity is preloaded on the device. Some systems generate device identities on the fly. These identities are either temporary or be written back to the devices as permanent identifiers. The provisioning service provides the device credentials to connect to the final system (such as a management system or a messaging system). In a large-scale system, devices are often partitioned into different partitions or tenants.

4. Acquiring software packages, patches and configurations

A device may need to be updated before it can work properly with the backend service, which may have gone through protocol breaking changes since the device was shipped. Hence, as the last step of provisioning, a device needs to download necessary software packages, patches and configurations. These bits can be acquired from the backend system, a package repository or from the provisioning service.

The bootstrapper configures the software package as an auto-start process when the device is powered up. At this point, the bootstrap's work is done, and the device is ready to be connected to the backend system.

8.3.2 Design a generic solution

In some cases, it would be desirable for a device to have the ability to attach to multiple backends, such as different cloud platforms or different on-premises management systems. Even if we could load multiple platform-specific bootstrappers to the device, we can't determine which bootstrap should be launched when the device is powered on because we can't predict the deployed environment. So, we need a bootstrapper for bootstrappers. COA's proxy bootstrapper can serve for this purpose. The bootstrapper connects to a single well-known discovery service. The discover service uses some means, such as user input, to establish which backend to use. Then, it returns a recipe to acquire the corresponding bootstrapper. Once the platform-specific bootstrapper is acquired and launched, the device can follow the aforementioned workflow to complete the provisioning process.

For large-scale deployment, manually selecting the backend system is troublesome and error-prone. One way to solve this problem is to configure all proxy bootstrappers to prefer a local discovery service. Then, launch any of the devices. Finish backend selection and then allow the device to acquire capability to act as a discovery service that points to the selected backend platform. Now, turn on the rest of the devices. Because the proxy bootstrappers are configured to prefer local discovery service, they will use the discovery service hosted by the first device and connect directly to the selected platform.

It's also possible to use this process to connect to multiple platforms simultaneously. For example, a device can be managed by one platform while taking workloads scheduled by another system. And the device can use a message pipeline created by a third system. Finally, after the device is provisioned, the capability proxy can be used to apply updates and patches as well.

8.4 COLLABORATIVE COMPUTING

When I grew up, "multi-player" meant to use two joysticks attached to the same Nintendo console. Although networked multiplayer games did exist, they hadn't gain popularity in China at the time. My personal experience with multiplayer games started with games like Command and Conquer and Quake around the time I was in college. Multiplayer games are exciting because your teammates or opponents are controlled by other players, who are much more intelligent than machine-controlled bots. Although modern games can use AI-driven bots to provide much sophisticated gameplays, they still can't compete with creative, sly humans.

With the rapid development of Internet, traditional single-user activities are becoming collaborative. The most obvious example is collaborative document editing with Google Docs or Microsoft 365. Microsoft's Fluid framework breaks the boundaries among applications and allows you to leverage all Office capabilities in context without switching applications. Coincidentally, Fluid framework to certain extent is the first production-grade application of COA. You can see some COA concepts such as capability discovery and delegate resonating well with Fluid design.

A key aspect of enabling collaborative computing is to synchronize states of multiple participating parties. Figure 8.4 illustrates some of the approaches to synchronize states: (a) Uses a shared state store. Both users directly update the shared store. This is a strongly consistent scheme.

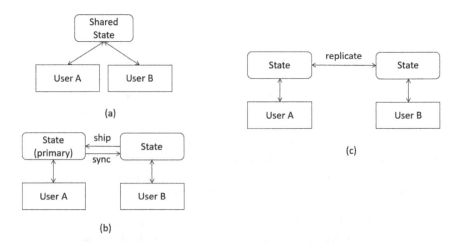

Figure 8.4 State synchronization schemes.

Concurrent updates are sequentialized by locks or accepted/rejected based on Etag matching. The downside of this approach is latency and potential bottleneck and deadlocks. (b) It elects one of the parties as the primary. All state updates are shipped and reconciled on the primary state store. And the reconciled state is synced back to secondaries. Due to the extra latency introduced by the ship-sync process, this solution doesn't work well with highly interactive scenarios in which frequent data exchanges. It's also hard to implement reliable primary election that tolerates network partitions and (c) uses a local store for each party. And the local stores are constantly synchronized and reconciled. This is an eventual consistent scheme that ensures all user states are eventually synced. Writing an efficient, scalable and reliable data replication system is one of the most difficult problems in distributed computing. We suggest you investigate existing solutions instead of trying to create a new one. What you create is likely to work well in limited scenarios. But you'll face very illusive and hard to fix problems when the system is under stress and when the network condition fluctuates. Frameworks like Microsoft Fluid provide a solid foundation for frequent multi-party collaboration. Leveraging such as framework would yield much better returns of your investments.

In both multiplayer gaming scenario and collaborative editing scenario, a general assumption is that the participants are collaborative, which means they sincerely intent to achieve something together. However, the Internet as we know it is a hostile environment. We can't always safely assume every party has good intentions. It's an interesting challenge to allow legit users to safely work with each other while detecting and deflecting attacks from untrusted parties and unexpected adversaries. For example, a hacker may attack a smart power grid by installing devices that interfere with how smart meters report usage data back to the network. This leads to faulty reports that are lower than actual usage. One way to solve this is to place a few nearby smart meters into a small consortium that runs a consensus algorithm to cross-validate usage data before the data is reported upstream.

To accommodate for collaborative computing scenarios, we extend COA to include a concept of *collaborative capability*. In addition to make offers as a sole capability provider, a device or service can make offers to participate in delivering a collaborative capability. A COA implementation may provide leader election mechanism so that a lead capability provider can be appointed as the primary point of capability delivery. For example, in a multiple player scenario or a voice conference scenario, one device can be elected as the lead server serving for the group. And when the lead server fails, the collaborative capability offer becomes invalid and a new round of discovery processes is launched to establish a new lead.

Concurrent mutations on the same data are the biggest challenge to collaborative computing. *Event Sourcing* (ES) is an architecture pattern that can be used to implement synchronizations among multiple parties in collaborative computing. In an ES system, all transactions are recorded as events. You can start with an initial state and play back all recorded events to get the latest state, or to play back events to a certain time to get the state of the time. Further, if we can guarantee all events from all parties are well ordered, we can use a simple *last-write-win* policy to resolve conflicts, especially when we reduce the size of objects to be synced – such as a sentence instead of a paragraph. Under this design, all parties publish their events, which are ordered and synced to all other parties. Then, the events are applied in order with last-write-win conflict resolution policy. This is an eventual consistent model, in which participant states will eventually be in sync. If a bigger object (such as a document) can be segregated into smaller objects, we can run synchronization upon each object in parallel, providing a real-time collaboration experience.

8.5 CONTEXT-AWARE COMPUTING

We briefly discussed device context in the last chapter. Making a device adaptive to the surrounding environment creates some of the most challenging computer science problems. It also creates some of the most exciting opportunities for edge computing.

Many of robotic system problems are concerned with understanding and reacting to the surrounding environments. For example, a robot like Boston Dynamics' Spot robot dog needs to be able to navigate through different terrains around various obstacles such as stairs, pits and rocks. A self-driving car needs to constantly survey its surroundings and make rapid decisions to react to dynamic road conditions. A drone in a hive needs to run local collision avoidance algorithms to avoid colliding into other drones. COA can be extended to provide structured device context descriptions so that a device can dynamically discover and acquire capabilities to react to environmental changes. We call this *device context descriptor.*

8.5.1 Device context descriptor

When you accidentally touch a hot object, you automatically withdraw your hand from it. This is called the *withdraw reflex*, which protects you from damaging stimuli. This is controlled by a neural pathway named *reflex arc*, which triggers the avoidance action without going through your brain. We also demonstrate other types of reflexes that help us to avoid various hazardous conditions, such as sneezing that gets rid of foreign particles from nasal cavity, muscular defense that contracts abdominal muscles to shield against a blow on the stomach and a less obvious baroreflex that stabilizes body blood pressure.

It's easy to design an artificial reflective system. For example, we can set the safe temperature range and trigger an avoidance action when the temperature exceeds a threshold. Then, we access the effectiveness of the action by validating if the temperature has fallen back into the safe range. Under COA, a device can be associated with several attributes with defined ranges. When an attribute exceeds the defined range, the device uses its built-in capabilities, or discovers a new capability that can bring the attribute back to the safe range.

The artificial reflective system can be used not only to avoid harmful conditions, but also to seek preferable conditions. For example, a mobilized device can roam around to locate better Wi-Fi signal strength.

The reactive system allows devices to generate actions to adjust for condition changes. As we mentioned in the last chapter, we tentatively call this *artificial motivation*, as it simulates a living entity being motivated to take various action in reaction to its environment. Device context descriptor provides a generic and simple mechanism for a device to perform corrective actions to maintain a set of attributes within a desired range. A device can also exchange descriptor data with nearby devices so that the group is aware of not only the environment but also conditions of other devices.

8.5.2 Adapt to network conditions

In Chapter 7, we described how capability proxy can switch between a locally hosted capability and a cloud hosted capability based on connectivity to Internet. And we've also introduced how a capability proxy can adjust its behavior based on network conditions – such as sending low-resolution data when the bandwidth is low, and switching to high-resolution data when additional bandwidth is available. In both situations, we assume there's a network application programming interface (API) that the capability proxy can invoke to reach network conditions. This section offers a brief discussion on how the network API may look, and how you can achieve different scenarios with the help of such an API.

We observe there are three tiers of network APIs, especially in a 5G network API: query, react and manipulate.

- Query

At this most basic tier, a network API offers routes for applications to query network conditions, such as available bandwidth, average latency and cell tower congestion level. The API can also serve device-specific queries, such as listing available edge sites based on a mobile device's identity and geographic location.

- React

The second tier of the API offers support for event-driven programming model by publishing network events to a messaging backbone. Applications subscribe to selected topics and implement logic to respond to network events.

- Manipulate

The third tier is the most exciting tier. This tier uplifts the network from an infrastructural component to an intelligent entity that can be factored into the application design. Through this tier of API, your applications can influent how network behaves. For example, you can dynamically request more 5G bandwidth allocations or adjust network packet sizes to fit your scenarios.

We can take different approaches to incorporate network APIs into applications. The most straightforward method is to expose network API as a capability that applications can consume. To the other end of the spectrum, we can make the network API complete transparent to applications (and even to capability proxies) if we integrate network API directly into a *service mesh* that is designed to provide software-defined networking layer for applications to communicate with each other. For example, the service mesh can implement dynamic routing rules that transit between different connections based on network API telemetries.

8.5.3 Real-world contexts

As we stated earlier, edge computing is *computing in context*. When the cyber world and physical world collide, exciting scenarios arise. With developments of sensor technologies, devices have abundant opportunities to understand and react to real-world conditions, such as lighting, temperature and humidity. A device can also acquire knowledge about its own condition, including acceleration, orientation, geographic location and many others. Furthermore, with AI technologies, a device has unprecedented opportunities to understand the world around it.

COA's common lexicon provides a mechanism to assign real-world features with common names that can be shared and understood among all COA applications. The value of a primitive is a primitive type, such as an integer, a decimal or a string. A value is associated with a unit, which is also described by a primitive noun in the lexicon. A more complex structure is decomposed into primitive values. And it can be associated with a *projection* that describes how data is encoded and serialized for transportation, such as a JOSN document serialized as byte arrays. For example, a temperature value can be described by the following quadruple:

Temperature, 68.0, decimal, Fahrenheit

And a location can be described by the following octuple:

Latitude, 37.0, decimal, WGS84, Longitude, –122.0, decimal, WGS84

When multiple devices publish such data to the network, a compute unit can build up a comprehensive map of the reality around it and perform intelligent actions. For example, a datacenter monitor system can build up a heat map of server aisles and use the map as a guide to optimize airflow in the datacenter to achieve more efficient cooling.

This leads to another COA concept that we haven't mentioned before – a *common data box*. The common data box allows participants to publish and receive data described with COA lexicon. By tapping into this database, a device can gain additional insights into the environment and implement logic to adapt to environmental changes. Consider the following scenario: a surveillance camera in a meeting room periodically publishes captured images to the box. A smart conference system detects the availability of the images and uses facial recognition to recognize who's in the meeting room. Then, it compares with its calendar and automatically launches the meeting when the person steps into the meeting room. When there are multiple cameras reporting data, it can switch between different feeds to give participants better views when different people speak.

A data item in the common data box can be associated with a location and a valid time. This means the data can only be picked up by devices within a certain range during a specified period. This association with real-world location and time enables many very interesting scenarios. We'll offer a few examples below as food for thought:

- A store pins a digital coupon within a 5-mile range around the store. When a mobile user is nearby, they can use their cell phone to pick up the coupon. As they are already near the store, the probability that they are motivated to visit the store is much higher than seeing the same coupon in their mailbox when they are at home. The coupons can also be associated with more aggressive expiration times (such as in hours) to further boost the conversion rate.
- A college professor sets up a virtual beacon in the classroom. All students need to check in with their credentials during the class time. A group of travelers can also set up virtual beacons at rally points to make sure everyone is accounted for. Museums and parks can pin informational content – such as video and audio clips – so that tourists can access more background information during their visits. Another possible usage of a virtual beacon is to set up virtual roadblocks to block damaged roads so that GPS systems can plan routes around the obstacles. Of course, in such cases the virtual beacon needs to be secured to avoid attacks.
- An individual can pin virtual time capsules at certain locations and configure them to be accessible in 30 years. When the time comes, people visiting the locations can pick the time capsules to reveal the contents. A time capsule can contain monetary value – such as access to a certain amount of digital currency. This is a nice way to pass down some wisdom and fortune.
- Users can pin arbitrary messages at any locations – over the lake, at mountain top, in the ocean – so that others can pick up. As you travel

around, it is a nice experience to pick up messages left by earlier visitors. This also gives people a channel to leave messages at points of interests without carving physical marks into priceless historical landmarks and artifacts. A message can be permanently pinned or be picked up and carried to a new location. This enables additional scenarios such as a virtual Olympic torch relay in which many virtual torches can be carried around the world. It's exciting to imagine the unprecedented level of participation to spread the Olympic spirit farther than ever before.

- A social news website can assign very short expiration times – say 15 minutes – to user-posted messages. When a viewer clicks to "like" a message, the message's lifetime is extended by a few seconds. This design ensures only the most recent and exciting news messages are kept on the page. And in addition to "like" or "dislike", users can also append additional information to either support or refute the original story. Then, the website can stick scattered information pieces into a complete timeline that presents views from both sides.

There are many more exciting scenarios we can think of when we blend the cyber world and reality. And this is what COA is about – to allow users to stay in their real-world contexts and consume any capabilities the cyber world may offer without thinking about switching contexts among applications and services.

8.6   APPROACHABLE AI

Another use case of COA is to make AI approachable. Wrapping up an AI model for easy inference is not particularly hard. An AI service provider can wrap a trained AI model as a REST endpoint, and user triggers a new inference by posting user data to the endpoint. In theory, we can also containerize an AI model so that it can be pulled down and provider inference service on the edge in disconnected scenarios.

8.6.1   AI inferences

Providing offline inference faces some practical challenges. Below we summarize some of the challenges and introduce how COA constructs can help to address them:

- Sophisticated AI models are often quite sizable, making them unsuitable for low-power devices. COA's acquisition mechanism can be used to acquire the AI model on edge. And when the model can't be fit into an edge device, COA's delegation mechanism can be used to delegate AI inference calls to a more capable edge device such as a local server acting as a filed gateway.

- Edge AI models may need to be updated. COA's acquisition mechanism can be used to automatically detect and download updated AI models. And while the local AI models are being updated, the capability proxy can temporarily switch to the cloud-hosted AI model to provide continuous service while the local model is being updated.
- When the AI model runs on the edge, we need to put associated offline billing support in place to continue charging for consumption. COA can help to bring an immutable ledger to edge to keep track of consumption and reconcile with the server when the connection is restored.

COA capability proxy's middleware mechanism can also provide supports to online inference as well, such as providing authentication, secured connection, batching, compression and encryption. You may ask if AI model can still run inference if the input data is encrypted. As a matter of fact, privacy-preserving AI has become a hot topic of research in the past few years, and it's expected to have growing importance in the future, especially when more and more data private legislations like EU's General Data Protection Regulation are put in effect. Fortunately, researchers have come up with various ideas to protect privacy during training as well as inference. For instance, *homomorphic encryption* allows training and inferences to be carried out directly on encrypted data. And secured multi-party computation allows multiple parties to collectively calculate results while keeping the inputs private. COA's middleware can be used to hide the complexities of these systems and provide a secured inference system to the broader developer community.

8.6.2 Custom AI models

Data plays a pivotal role in quality AI models. Creating a robust, generic AI model requires not only the genius of scientists, but also huge amount of quality data. However, not everyone can afford to collect high-quality data sets. Custom AI models provide a practical solution for specific scenarios using much smaller data sets. If we train an AI model using the specific data from a scenario, the model is heavily biased (or called "overfit") toward the trained situations. Although the model is unlikely to be generalized for other situations, it can provide a very satisfactory result. For example, a custom voice recognition model can be trained to adapt to the specific ambient sound patterns (such as sound of running machinery) to provide more reliable detection than what can be provided by a generic model.

COA's capability proxy is by design not just a passive proxy to a web service. Instead, it's ultimately responsible to deliver high-quality capability to consumer applications. Hence, it shall implement logics not only to locate and acquire the best possible solutions, but also to improve existing offers so that it can provide better options to the requestors in the future. In this case, the proxy can be extended to provide feedback to the capability provider. It can feed the request data to a training endpoint to provide additional training data to customize the AI model for the specific scenario. The training process can happen on cloud or be delegated to a capable computer (with a powerful graphics processing unit [GPU], for instance) on the local network. Then, the trained model can be shared with other clients through data sync.

When conditions permit, the capability proxy can also run background optimization processes independent from user requests. For instance, the proxy can drive a hyper parameter search process to optimize AI model settings and share the findings with other clients. A coordinated search is also possible, with search space segmented and distributed among available clients. Certain automatic, fine-granular modifications to the model structure itself could also be made feasible, such as picking different neuron types to represent different interaction patterns among key attributes.

8.7 BROWSER-BASED CLOUD

We've discussed the idea of an *edge cloud* that is operated by telecom carriers using their pervasive device fleet deployed to regular homes. We've gone through many crazy ideas in the past two chapters. We'd like to close this chapter (and this book) with a final reckless idea – a browser-based cloud.

Browsers are easily most pervasive distributed compute engine in modern world. And with the increasing emphasis on security, they offer safe sandboxed execution environments that are well isolated from the host environments. At any given moment, there are millions of browser instances running. This gives us a sizable, distributed compute environment that is literally always on and always available. At the same time, for most users, browsers are the most common way to consume various cloud services – social networks, online productivity products, file storage, emails, online banking – in pretty much all aspects of our lives nowadays. This correlation between compute and consumption of computation results gives us a unique opportunity to create a cloud platform that offers compute capabilities right at where these capabilities to be consumed. We envision that in many cases the compute invocations never even leave the machine boundary, with the best latency we can hope for.

Each browser on the network is both a consumer and a provider. It offers certain compute resources to the network, and it consumes compute power offered by others to complete complex tasks. Additional delegated edge servers (such as those devices owned by the telecom companies) can be added to the network to boost performance of busy segments. And a user can also use additional local machines (such as idle laptops) to boost workloads in here local networks. Figure 8.5 presents a low-fidelity mockup of how such a cloud platform looks like in a modern browser. With more modern companies switching to a pure online working environment, such browser-based cloud platforms have great potentials to provide and end-to-end, high performance online office environment that supports the complete spectrum of activities, from design to develop to test to release.

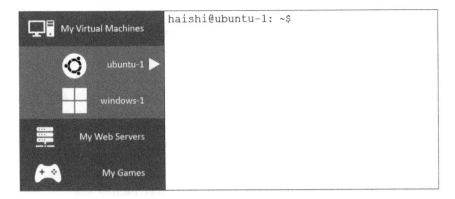

Figure 8.5 A browser-based cloud.

And this browser-based experience is not a secondary approach to access native systems – it is the system.

To join a browser to the compute network, the user just needs to launch a browser and navigate to a designated website. And to leave the network, she can simply shut down the browser. This forms a very dynamic compute plane. We need some schedulers such as the quicksilver scheduler we mentioned in Chapter 7 to handle workload scheduling on such a compute plane. We'll also need to face some other challenges, such as establishing and isolating secured virtual compute tenants, protecting user files and resource metering and billing, etc.

To host arbitrary workloads on browsers, we can consider implementing a container runtime on top of WebAssembly. This will enable us to run legacy Docker images (it's interesting to think about how we refer Docker images as "legacy" in this context) natively on browsers. As WebAssembly continues to mature, we are gaining increasing flexibility and compatibility to host advanced workloads that use threads, GPU acceleration and single instructions multiple data (SIMD). And we can use out-of-browser WebAssembly runtimes to further extend the compute plane to low-power edge devices.

Consider the online game streaming scenario – with browser-based cloud, an online game service provider can set up dynamic local area network (LAN)-based gaming networks for smaller group of players to play in game sessions hosted by a locally elected server. This is a more scalable and more economical approach than building up huge gaming centers with hundreds of thousands of game servers. What's even more interesting is that the existing gaming consoles can also be recruited into this compute plane to augment gaming experience by serving as secured local file cache to reduce game package transfer time, hosting AI-driven bots to either assist players (such as driving assistance in driving simulation games) or challenge players with autonomous opponents. Although we have access to more bandwidths and more compute power nowadays, the demand of enabling new gaming scenarios such as high-resolution VR/AR and hologram scenarios keeps growing. When we can recruit as much local compute resources as possible, we can push the boundary to enable additional gaming experiences.

The browser-based cloud can be consumed by browser-only devices. This allows us to create various low-cost education, productivity and entertainment devices that bring the complete cloud experiences to a broader audience. With COA, we can cache as much capability as possible on client devices, or to a few more capable devices that the low-power devices can reach. This enables offline or sparsely connected scenarios such as a classroom in rural areas, or a squad in a jungle. As a matter of fact, a user doesn't even need to carry a device around. She can access all here contents using any public or shared devices, given a secured private session can be established.

8.8 COA – A CONFESSION

We've rambled on for two chapters on the topic of COA, because we believe it's a worthy topic for a thorough discussion. We certainly don't claim COA is a brand-new invention. Instead, COA is a combination of observations of what others are doing and some new concepts that we envisioned to enable the new architecture paradigm. We must admit that because COA is still being conceptualized, there are many missing details and loose ends to be tighten up. But we think we are onto something. And we hope these chapters will inspire some thoughts to further develop (or to contradict with) COA. This is because we believe edge is not just the edge of cloud. Edge computing has its own unique characteristics that deserve the community to think of "edge native" application designs so that we can create more efficient and robust edge solutions.

We think the capability proxy is the most fundamental and enabling piece of COA; semantic service discovery is a solid attempt to generalize COA; and quicksilver is a key idea to bring COA to the edge scale. These are all grand ideas that are mostly on paper at the time of writing. It takes much effort to implement these ideas, and we hope we can help as much as we can to pave the path for these ideas to become a reality.

8.9 SUMMARY

COA has many interesting revolutionary application scenarios, such as breaking down the boundaries of applications, building intelligent applications that can dynamically acquire new capabilities, creating context-aware and adaptive applications, making AI approachable to the whole community and building a browser native cloud platform. COA is still a new idea that needs further development. We hope we can bring COA to reality with the help of the whole community.

Index